Cover photo courtesy of Mark Keller.

Published by
Charles E. Merrill Publishing Company
A Bell & Howell Company
Columbus, Ohio 43216

3.4 (3-21) From W. Bascom, *Waves and Beaches: The Dynamics of the Ocean Surface,* copyright ©1964 by Educational Services Incorporated. Reprinted by permission of Doubleday & Company, Inc.

This book was set in Helios.
The Production Editor was Linda M. Johnstone.
The cover was designed by Will Chenoweth.

International Standard Book Number: 0–675–08576–4
Library of Congress Catalog Card Number: 76–2959
1 2 3 4 5 6 7 8 9 10—83 82 81 80 79 78 77 76

Printed in the United States of America

preface

The Ocean World presents oceanographic science in a simplified and graphic form. Most of us approach the ocean with a mixture of reactions, ranging from wonder and awe to dread of an unfamiliar and seemingly hostile environment. In this book, we seek to make the ocean and its processes more familiar and to illustrate the forces that drive it—from the waves at the surface to the circulation of deep-ocean waters.

We present oceanography as an interdisciplinary modern science in which geologists, chemists, physicists, and biologists are drawn together to study the processes that control major features of the ocean basins, the composition of seawater, and the ocean currents.

Ocean science has advanced dramatically in the past few decades. Some progress has been made because of the increased resources devoted to learning about the ocean. Other advances have come from the use of new instruments for measuring ocean processes, and still others come from the satellites and submersibles that permit us to observe directly the ocean surface and ocean bottom.

The ocean is becoming more important to us because of our increasing dependence on its waters and its floor as sources of protein and of petroleum and natural gas to fuel the world's industrial economy. In the process of exploiting ocean resources, we need to protect the ocean against irreparable damage through overfishing or pollution. These various aspects of the ocean-dominated world are explored in this book.

The Ocean World is divided into four modules. Module 1 deals with geological oceanography and describes the basin that contains our world ocean. Module 2 explains ocean circulation and shows how physical oceanographic processes control global distribution of heat from the sun. Waves and tides, which constantly smooth and shape continental margins, are dealt with in Module 3. In Module 4, the coastal ocean, where humans interact with marine environments, is covered. It is in the coastal ocean that ocean resources are most valuable to us, and potential losses due to unregulated exploitation greatest.

In presenting this view of the ocean world, we outline the major concepts of marine science in an audiovisual format. Then, we summarize the material using the more traditional book form so that students have an opportunity to review major concepts and supplement them with concrete details. Outlines, exercises, questions, and other special aids reinforce the student's grasp of scientific principles and heighten interest in this most fascinating branch of earth science.

to the student

This program is designed to introduce you to oceanography using a variety of media. MEDIAPAK audiovisual presentations are integrated with this textbook and can be used throughout the course to enable you to master concepts of oceanography faster and more thoroughly than if you used a conventional text alone.

FORMAT OF THE TEXT

This program is divided into four separate units, or modules. Each module contains the following:

Introduction—A brief look at the material to be covered

Learning Objectives—Statements of what concepts should be learned

Key Terms—A list of difficult terms, and words that might be unfamiliar to you

Outline—An outline of the audiovisual presentation, which will be helpful for reference and review

Exercises—Questions and problems designed to reinforce your learning. The answers to these self-evaluation exercises appear in the back of the text

Summary—A review of the audiovisual presentation; contains a number of the key visuals from the audiovisual materials for further study

Questions—Problems designed to help you probe more deeply into the concepts presented and to relate these concepts to ideas presented in other modules

Suggested Readings—A list of additional available sources of the topics covered

The textbook also features a glossary, which contains definitions of all Key Terms and other important concepts.

PROCEDURES

Before going to the audiovisual presentation for a given module, you should read the corresponding introduction and study the Learning Objectives and Key Terms. Then, select the MEDIAPAK component(s) for that module. Note that each visual frame contains a frame number, which corresponds to a frame announcement. You will hear the frame announcement immediately preceding the audio description of the visual. You may stop your audiovisual equipment if you need more time on a particular frame. Also, if you do not understand a concept, you may rewind the equipment and review the information. After completing the audiovisual portion of the module, return to the textbook and answer the Exercises. Answers are provided in the back of the book. You may read the Summary for review and further study of the material covered in the audiovisual sequence.

credits

Book

I.1, 1.9, 1.12, 4.5, 4.11 left NASA. **I.2** Reproduced from the collection of the Library of Congress. **I.4** Plessey Environmental Systems. **1.3, 2.6, 2.7, 2.9** Goode's Base Map Series. Copyright by the University of Chicago Department of Geography. **1.3** U.S. Naval Oceanographic Office, "Chart of the World," 10th ed., H.O. 1262A. **1.6** Painting by Heinrich Berann, courtesy of Aluminum Corporation of America. **2.4** U.S. Naval Oceanographic Office, *Instruction Manual for Oceanographic Observations,* H.O. Pub. 607, p. 42. **2.6** H.J. McLellan, *Elements of Physical Oceanography* (Pergamon Press, 1965), p. 44. **2.7** H.U. Sverdrup, M.W. Johnson, and R.H. Fleming, *The Oceans: Their Physics, Chemistry, and General Biology* (Prentice-Hall, 1942), Chart VI. **2.9** U.S. Naval Oceanographic Office, "Pilot Charts," various printings, and *Encyclopaedia Britannica World Atlas* (1955). **2.10** U.S. Naval Oceanographic Office, *The Gulf Stream* 6. **2.11, 4.3, 4.6 right, 4.10** M. Grant Gross, *Oceanography* (Prentice-Hall, 1972). **3.4** W. Bascom, *Waves and Beaches: The Dynamics of the Ocean Surface* (Anchor Books, Doubleday, 1964), p. 86. Copyright 1964 by Educational Services. Reprinted by permission of Doubleday and Company, Inc. **3.6** U.S. Naval Oceanographic Office Pub. 700, Sect. 1, 1968. **3.7** U.S. Coast and Geodetic Survey, *Tidal Current Charts: Long Island Sound and Block Island Sound,* 4th ed., Serial 574, 1958. **3.8** H.A. Marmer, *The Tide* (Appleton-Century-Crofts, 1926). **3.10** W.E. Yasso, and E.M. Hartman, Jr., *Beach Forms and Coastal Processes,* New York Sea Grant Institute, 1975. **4.4** P. Groen, *The Waters of the Sea* (Van Nostrand, 1967). **4.6** D.W. Pritchard, *American Society of Civil Engineers Proceedings* 81. **4.8** J.P. Tully, and A.J. Dodimead, *Journal of the Fisheries Research Board of Canada* 14: 241–319.

Media

1-title, 1-1, 1-19, 1-22, 1-28, 1-31, 2-title, 3-30, 4-title, 4-1, 4-4, 4-5, 4-10, 4-16, 4-21, 4-22, 4-28 left NASA. **1-4** D. Ericson and G. Wallin, *The Everchanging Sea* (Knopf, 1967). **1-5** A.F. Spilhaus, *Geographical Review* 32: 434. **1-6** Data from G. Wüst, W. Brogmus, and E. Noodt, *Kieler Meeresforschungen* Band X: 139. **1-9, 2-12, 2-15, 2-19, 2-22** Goode's Base Map Series. Copyright by the University of Chicago Department of Geography. **1-9** U.S. Naval Oceanographic Office, "Chart of the World," 10th ed., H.O. 1262A. **1-14** Painting by Heinrich Berann, courtesy of the Aluminum Corporation of America. **1-20, 2-27, 4-7, 4-9, 4-12, 4-13, 4-19** M. Grant Gross, *Oceanography* (Prentice-Hall, 1972). **1-32** Courtesy of The American Museum of Natural History. **1-33** Deep-Sea Drilling Project, Scripps Institution of Oceanography. **1-34–1-37** NSF. **1-38** U.S. Coast and Geodetic Survey. **2-5** Plessey Environmental Systems. **2-8** U.S. Naval Oceanographic Office, *Instruction Manual for Oceanographic Observations,* H.O. Pub. 607, p. 42. **2-11** Heat budget after H.G. Houghton, *Journal of Meteorology* 11:7. Ocean-surface temperature data from W.E. Forsythe, ed., *Smithsonian Physical Tables,* 9th rev. ed. (Smithsonian Institution, 1964), p. 726. **2-12** H.J. McLellan, *Elements of Physical Oceanography* (Pergamon Press, 1965), p. 44. **2-13** Data from G. Wüst, W. Brogmus, and E. Noodt, *Kieler Meeresforschungen* Band V:146. **2-15** H.U. Sverdrup, M.W. Johnson, and R.H. Fleming, *The Oceans: Their Physics, Chemistry, and General Biology* (Prentice-Hall, 1942), Chart VI. **2-16** R.B. Montgomery, *Deep-Sea Research* 5:144. **2-18** J.P. Tully, *Journal of the Fisheries Research Board of Canada* 21:942. **2-19** U.S. Naval Oceanographic Office, "Pilot Charts," various printings, and *Encyclopaedia Britannica World Atlas* (1955). **2-21** R.H. Fleming, *Geological Society of America Memoir* 67:95. **2-25** National Environmental Satellite Service. **2-26** U.S. Naval Oceanographic Office, *The Gulf Stream* 6. **2-28** A.N. Strahler, *Physical Geography,* 2d ed. (Wiley, 1960), p. 129. **3-1** Betty Ann Morse. **3-3, 3-4**

Portland Cement Association. **3-9, 3-11–3-14** Coastal Engineering Research Center. **3-17, 3-18** NOAA. **3-20, 3-21** W. Bascom, *Waves and Beaches: The Dynamics of the Ocean Surface* (Anchor Books, Doubleday, 1964), p. 86. Copyright 1964 by Educational Services. Reprinted by permission of Doubleday and Company, Inc. **3-26** U.S. Naval Oceanographic Office, *American Practical Navigator,* H.O. Pub. 9, p. 712. **3-27** U.S. Naval Oceanographic Office Pub. 700, Sect. 1, 1968. **3-28** U.S. Coast and Geodetic Survey, *Tidal Current Charts: Long Island Sound and Block Island Sound,* 4th ed., Serial 574, 1958. **3-29** H.A. Marmer, *The Tide* (Appleton-Century-Crofts, 1926). **3-32** W.E. Yasso, and E.M. Hartman, Jr., *Beach Forms and Coastal Processes,* New York Sea Grant Institute, 1975. **4-2** E.C. LaFond, *Scientific Monthly* 78:243–53. **4-8** P. Groen, *The Waters of the Sea* (Van Nostrand, 1967). **4-11** D.W. Pritchard, *American Society of Civil Engineers Proceedings* 81. **4-15** J.P. Tully, and A.J. Dodimead, *Journal of the Fisheries Research Board of Canada* 14:241–319. **4-33** Conoco. **4-34** Salt Institute, Alexandria, Virginia.

contents

introduction

Oceanography is the scientific study of the earth's most distinctive feature—the ocean. Satellites show the ocean dominating the earth by covering 71 percent of its surface. Even the clouds that obscure half the planet came from the ocean (Figure I.1).

Figure I.1 *The earth's Western Hemisphere, as photographed from* Apollo 8.

In these four modules, we examine an ocean-dominated world in which the continents are viewed as barriers to ocean currents and sources of much of the sediment that blankets the ocean floor. We hope to show you how important the ocean is to your life, and how it shapes the earth and influences the behavior of the atmosphere.

Study of the ocean has a long history, most of it related to solving very practical problems. The earliest ocean exploration was related to the need for new trade routes, followed by exploration of the New World for exploitation by Europeans. During these early years of ocean discovery, the boundaries of ocean basins were explored. But precise location and mapping of the shorelines had to wait until navigation was improved through the development of accurate timekeeping aboard ship. This problem was not solved until 1761, when the chronometer, an accurate ship's clock, was developed by John Harrison, a British carpenter's son.

It was easy for the ancients to measure latitude, or distance north and south of the equator, because one can measure the elevation of the North Star above the horizon; it is directly overhead at the North Pole and on the horizon at the equator. But measuring east-west distances required accurate timekeeping at sea. Setting a clock at a given starting point at noon, when the sun is directly overhead, one can measure east-west distances by calculating the difference between the clock's time and the time of local noon. Since the sun travels 15 degrees per hour, time differences can be directly translated into longitudes, or distances east and west of a standard longitude, for instance that of Greenwich, England.

Mapping ocean currents was important for sailing ships. These slow-moving vessels made much better time crossing the ocean on their way to England when they used the strong Gulf Stream. Returning, they saved time by avoiding these currents (Figure I.2). Today's powerful ships are less affected by ocean currents. Now, currents are studied in connection with pollution control, fisheries management, and climate forecasting.

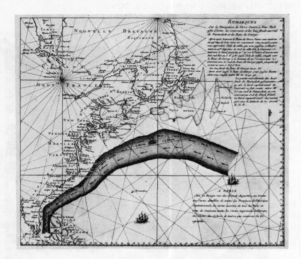

Figure I.2 *Map of the U.S. Atlantic coast, showing the path of the Gulf Stream, as reported by Benjamin Franklin.*

Detailed exploration of the ocean floors began with the laying of submarine cables during the mid-nineteenth century. The ocean depths were laboriously "sounded," first by weighted ropes, and later by piano wires lowered from ships. From these crude depth determinations, it became evident that ocean basins are not featureless plains but have large mountain ranges and narrow, deep *trenches* as rugged as anything on the continents.

The first systematic exploration of the ocean in all its aspects came between 1872 and 1876, when the *Challenger* expedition set out from England to collect samples and make observations in all parts of the ocean (Figure I.3). The 50-volume report of this expedition set the stage for the development of oceanographic science for more than half a century.

Oceanography has benefited from advances in related fields. For example, an important technical advance in ocean-basin exploration was the development of the sonic echo sounder. This device was developed to detect submarines during World War I. Sound pulses were bounced off objects near the sound source. The distance to the object was measured by timing the interval between sending the signal and receiving the return—hence the name echo sounder. In the 1920s, the first accurate and detailed maps of ocean-basin floors were made using this device. Mapping of ocean basins was greatly expanded by World War II, and exploration con-

Figure I.3 *H.M.S.* Challenger, *a converted British warship, cruised for four years on a nineteenth-century oceanographic expedition.*

tinues even today. The determination of ocean-basin shapes is an important objective of earth-orbiting satellites to be launched in the early 1980s. And locations of remote islands are still being adjusted as we learn more about the earth's shape.

Like all branches of science, study of the ocean is greatly dependent on instruments to make and record the necessary observations. For instance, temperature and *salinity* characterize water masses in the ocean much as temperature and humidity are used by meteorologists to track movements of air masses. The small temperature range in deep-ocean waters required precision measurements and necessitated the development of special thermometers to operate at the high pressures of the deep ocean. Precise chemical measures were originally used to determine the salinity of seawater. Now, these techniques have been largely replaced by electronic instruments that permit more closely spaced and more accurate measurements of temperature and salinity, as well as of current velocity (Figure I.4). As always, when a new observational technique is developed, our knowledge of ocean processes is expanded.

Oceanography is an example of "big science," requiring teams of scientists working together and expensive equipment to collect data. Originally, oceanographers were a few hardy individuals investigating questions of interest to themselves, often from small ships with only primitive equipment. Such studies could only deal with relatively small ocean areas.

The *Challenger* expedition showed the way for national expeditions exploring large ocean areas backed by the latest equipment. At first, these expeditions involved only one ship. With the development of new observing techniques, it became necessary to use multiple-ship expeditions to study well-selected processes. During the 1950s and 1960s, coordinated multiple-ship expeditions planned and carried out by international organizations provided the most dramatic advances and gave information about ocean areas that had previously been largely neglected, such as the Indian Ocean.

Development of earth-orbiting satellites and new observing techniques during the 1960s initiated a significant change in oceanography. These new platforms, with their capability for sensing ocean colors and

Figure I.4 *Electronic temperature-salinity-depth sensor going over the side of an oceanographic research ship.*

temperatures from great distances, permitted large-scale studies of major ocean processes. These latest techniques of observing from space are combined with more traditional means of observing the ocean by measuring currents from ships and buoys. The huge mass of data obtained from such studies is stored and analyzed through the use of computers, a task that would otherwise be nearly impossible.

In short, the study of the ocean has changed from individuals studying the ocean from ships to multiship surveys, and now to multidisciplinary studies in which oceanographers, meteorologists, and computer scientists combine forces to study ocean processes.

Advances in oceanography have had a major impact on other areas of science. For instance, development of new ideas about the nature of the earth's crust received powerful support from detailed studies of ocean basins after World War I. Detailed mapping of ocean sediments and geophysical properties of the *ocean basins* such as the magnetic field and gravity field over the ocean provided the raw material that was later synthesized into the concept of *sea-floor spreading,* which states that the earth's crust is formed in the middle of ocean basins and moves slowly toward deep trench areas, where it is destroyed. Continents and ocean basins are moved and reshaped by these processes.

In the 1960s, development of deep-ocean drilling capabilities enabled oceanographers to drill the deep-ocean floor to recover samples of the loose sediments and the underlying volcanic rocks. This technology has provided the raw materials for reconstructing the history of the development of the ocean basins. It has also provided data necessary for reconstructing changes in climate over the last million years of earth history.

All these projects are coordinated in the study of the ocean and show how solutions to problems posed by scientists can help answer questions of immediate concern to all of us—questions about development of mineral resources, pollution associated with waste-disposal operations, and prediction of climatic change.

MODULE

Geological Oceanography

INTRODUCTION

In this first module of our four-part oceanography survey, we deal with the earth's surface, its continents, and ocean basins. We have all seen a globe, which shows the earth as a huge sphere rotating on its slightly tilted axis. Here, the continents stand fixed in a flat, featureless ocean that is usually colored various shades of blue. Everything not clearly marked as land seems to be water. At first glance, we usually see only the large, complexly structured continental masses, interrupted by a more or less featureless ocean.

The goal of this first module is to cast some light through that blue curtain and perhaps to gain a new view of that part of the earth's surface covered by water. We will try to change our image by learning to see the earth as a vast, complex oceanic basin interrupted by continents, as well as by other important ocean-bottom features not visible to those who dwell on land. We will learn that the continents have not always had the same shapes, and that they have moved long distances under the influence of processes occurring in the nearly molten rocks beneath the ocean. Even now, they are moving and undergoing alterations in shape.

This view of the earth's crust as a continually changing structure has been accepted by most scientists only since the mid-1960s. But for as long as there have been accurate maps of the Atlantic Ocean, people have wondered why the Atlantic margins seem to fit like pieces of a jigsaw puzzle. Near the beginning of this century, Alfred Wegener, an explorer and meteorologist, formulated a theory that he called continental drift, in which he tried to show that continents move through ocean basins. But the ocean floor appeared to be, and in fact is, made of hard rock, and not of a pliable substance a continent could push through. Thus, there seemed to be no plausible mechanism by which continental drift could occur. However, the obvious fit of the continents continued to intrigue some scientists. Furthermore, the study of fossils showed that similar—even identical—ancient plants and animals once flourished simultaneously in areas now split by ocean basins. And mountain ranges and other land features that seem to have been at one time continuous are now separated by wide stretches of ocean. Without a plausible mechanism for continental transport, however, this evidence was not taken seriously by many scientists.

After World War II, a surge of interest in basic scientific research was matched by greatly increased support for research projects suggested by scientists themselves. New and sophisticated wartime techniques were adapted for studying continental and ocean-basin features. Among the earth features that could now be described in more sophisticated ways were the earth's magnetism, heat flow from the earth's interior, age and topography of ocean basins, and the distribution of earthquake belts. Data from these and other studies are best explained if one assumes that the ocean floor is not a continuous structure but exists as a series of separate, movable plates, whose shapes change with time. Volcanic rock is slowly but continuously added to one edge of a plate, while at a distant edge, the plate is drawn into the earth's interior and there consumed. Continents are passively carried along during this process, now colliding one against another, now being drawn apart or split down the middle.

These discoveries showed that two formerly mysterious topographic features of ocean basins play a role in the recycling of crustal material. The great, swelling mountain ranges that occur in each ocean basin were found to be regions where new ocean-basin material is generated by volcanic eruptions. Newly formed crustal rock is slowly moved down the sides of these vast submarine ridges and becomes part of the deeper ocean floor. At the narrow, curved *trenches* especially common near the margins of the Pacific Ocean, sea-floor material is destroyed by being drawn down into the earth's interior.

The dynamic picture of earth's crust that emerged from these studies has completely revolutionized earth science. For us, it provides the framework for our survey of marine processes. After studying the topographic features of ocean basins, their distribution, structure, and formation, we can then fill the basins with seawater, so to speak. In the later modules, we will see how the shape of ocean basins governs many of the dynamic processes that characterize the ocean itself.

OBJECTIVES

1. To become familiar with the locations of continental land masses, major ocean basins, and marginal seas on the earth's surface

2. To be able to identify major topographic features of the ocean floor, including continental margins, mid-oceanic ridges, and deep basins

3. To understand the theory of sea-floor spreading, and crustal plate movements, including formation of oceanic crust by volcanism at ridges, subduction at trenches, and faulting at other plate boundaries

4. To be able to explain the relationship of earthquake belts and volcanic sites to activity in the earth's mantle and movement of crustal plates

5. To understand the formation of coral reefs and atolls, and the relationship these structures bear to subsiding volcanic islands

6. To become familiar with the major sources of ocean sediment, their distribution, and transport

ocean basin
shoreline
continental margin
mid-oceanic ridge (rise)
trench
oceanic crust
continental crust
mantle
continental shelf
continental slope
continental rise
rift valley

crustal plate
transform fault
island arc
sea-floor spreading
gulf
reef (coral)
lagoon
fringing reef
barrier reef
atoll
core (sediment)
manganese nodules

KEY TERMS

Now you are ready to begin the audiovisual portion of this module. Select the MEDIAPAK 1 component(s) and proceed. Following is a topical outline of the audiovisual sequence. You will find this outline helpful for reference and review. After completing MEDIAPAK 1, return to this book to perform the exercises.

OUTLINE

Importance of the Ocean

Food, waste disposal, recreation, and climate modification

Extent of the ocean

The ocean and free water

The ocean and water from the earth's interior

Basic Features of the Ocean and Its Shores

The world ocean

Major basins, their boundaries, and distribution of land masses

Major features of the Pacific, Atlantic, and Indian oceans

Shorelines created by changes in sea level

Expansion or melting of continental glaciers

Effects of changing sea level on continental margins and coastal populations

Topography and Structure of the Earth's Crust

Continental crust, oceanic crust, and mid-oceanic ridges

Continental margins

Stable continental margins—shelf, slope, rise, and sediments

Unstable margins—trenches, mountains, and marginal basins

Structure of mid-oceanic ridges

Rift valleys and faults

Distribution of earthquakes and volcanoes

Deep-ocean basin on either side of a ridge

Crustal-plate Structure and Boundaries of Major Plates

Movement of crustal plates

Mountain building at regions of compression

Sea-floor spreading in the Atlantic Ocean

Characteristics of subduction zones—trenches, volcanoes, and island arcs

Characteristics of spreading sites

Oceanic-crust formation at ridges

Continental rifting

Age of continents (old) and ocean basins (younger)

Formation of Volcanic Islands and Island Chains at Centers of Volcanic Activity

Formation and Evolution of Coral Reefs and Atolls

Marine Sediments

Transport and deposition of lithogenous sediment

Biogenous sediment

Hydrogenous sediment

Dating of ocean-basin formation by the study of deep cores

Exercises

1. About two thirds of the earth's land area is in the ___Northern___ Hemisphere.

2. The three major ocean basins are connected around the continent of _____.

3. The ocean basin with the greatest north-south extent is the _P.O._.

4. *True or false:*
    A. Continental crust is destroyed by being drawn into trenches.
    B. The deepest earthquakes occur at sites of oceanic-crust formation.
    C. After volcanic islands form, they tend to sink into the ocean.

5. Atolls form when
    A. corals build a solid reef structure down the sides of a deeply submerged volcano.
    B. a fringing reef changes to a barrier reef as a volcanic island continues to grow.
    C. corals completely cover the reef structure with basaltic rock.
    D. an island subsides while the reef around it is built upward.

6. Iron-manganese nodules are
    A. derived from continental erosion.
    B. precipitated from seawater.
    C. common on continental shelves.
    D. dissolved at depths greater than 5 kilometers (3 miles).

7. Deep-sea drilling operations
    A. have been useful in dating ocean basins.
    B. recover long cores of continental crust.
    C. are carried on from platforms resting on the sea bottom.
    D. have shown that ocean basins are youngest near continental margins.

8. Match the following:
    A. Extensive glaciation causes
    B. Continental crust composed of
    C. Sites of crustal formation
    D. Sites of crustal destruction
    E. Oceanic crust composed of

    1. mid-oceanic ridges
    2. earthquakes and volcanoes
    3. transform faults
    4. granitic rocks
    5. lower sea level

F. Ocean basins formed by
G. Where crustal plates slip
     past each other
H. Unstable continental margins have
 I. Stable continental margins have
 J. Melting of continental ice causes

6. broad continental rises
7. sea-floor spreading
8. higher sea level
9. basaltic rocks
10. deep-ocean trenches

SUMMARY

Life on the earth is conditioned by the ocean. Most people live within a few hundred kilometers of its shores, so for them, it serves as recreational area, food source, waste-disposal site, and commercial highway. But even for those who live far from the ocean, it has a major influence on daily life. It stores the water necessary for life, and it absorbs and releases much of the sun's energy. This latter process creates the atmospheric circulation that causes our weather, and it prevents the extremes of heat and cold that characterize water-free planets.

Continents and Ocean Basins

The ocean covers nearly 71 percent of our planet's surface, to an average depth of 3,730 meters (12,200 feet). It hold more than 97 percent of earth's free water. A relatively small amount of water is locked in glaciers and other forms of continental ice, and an even smaller amount is contained in lakes, rivers, and the atmosphere. This water constantly evaporates from the ocean's surface and returns to it within a few years or less.

Throughout the four and a half billion years of earth's history, water has been released from the interior of the earth through volcanic action. Although a large, unknown quantity remains in the earth's interior, most of the water that has been released is now in the ocean. We know from the history of sedimentary-rock formation in water that an ocean has existed for at least three billion years.

Although the ocean is a single, interconnected body of water, for convenience we divide it into three major basins (Figure 1.1). Where no natural boundaries exist, as for instance around Antarctica, we draw arbitrary lines along meridians of longitude to separate the basins. The Pacific Ocean is bounded by the Aleutian Islands on the north, Indonesia and Australia to the east, and Antarctica to the south. The Atlantic Ocean includes the Arctic Sea to the north. The Indian Ocean extends from Africa and India to the Indonesian Islands.

Continents and ocean basins are not evenly distributed over the earth's surface (Table 1.1). About two thirds of the land is in the Northern Hemisphere. Even so, 61 percent of that hemisphere is covered by water. The Southern Hemisphere contains most of the world ocean. In fact, the ocean covers 81 percent of the earth's surface between latitudes 40 degrees and 65 degrees south. Here, there is almost no land to interfere with atmospheric or oceanic circulation.

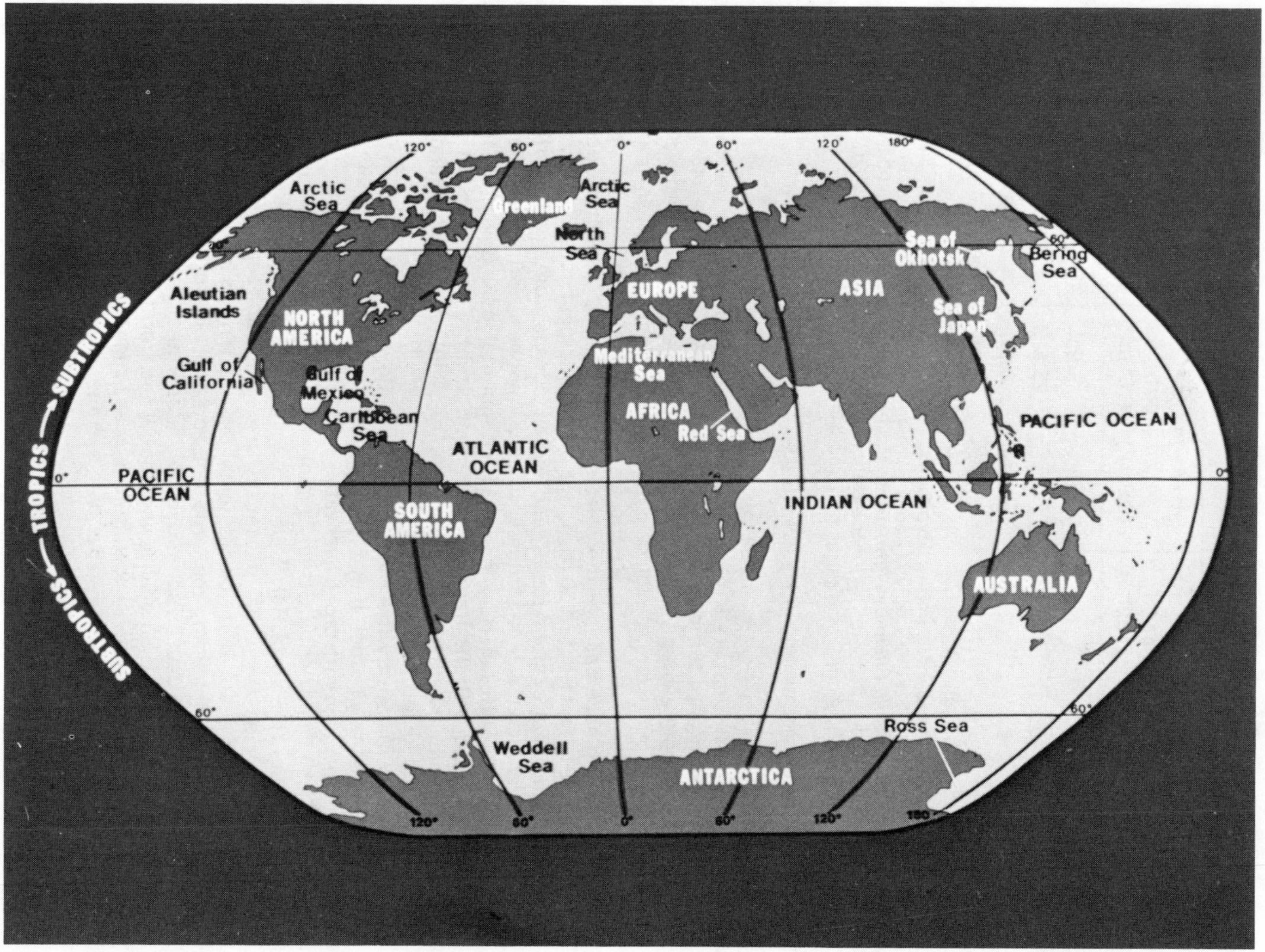

Figure 1.1 *The world ocean.*

TABLE 1.1   Areas and Depths of the World Ocean

Ocean area	Water area*		Land area drained†		Water / Land	Average depth	
	(10⁶ km²)	(10⁶ mi²)	(10⁶ km²)	(10⁶ mi²)		(m)	(mi)
Pacific	180	69.5	18	6.95	10	3,940	2.44
Atlantic	107	41.3	67	25.8	1.6	3,310	2.06
Indian	74	28.6	17	6.57	4.3	3,840	2.38
World ocean	361	139	102	39.4	3.6	3,730	2.32

SOURCE: H.W. Menard and S.M. Smith, "Hypsometry of Ocean-basin Provinces," *Journal of Geophysical Research* 71 (1966): 4305, and J. Lyman, "Chemical Considerations," in *Conference on Physical and Chemical Properties of Sea Water,* Publication 600 (Washington, D.C.: National Academy of Sciences-National Research Council, 1959), p. 89.
*Includes adjacent seas. Arctic, Mediterranean, and Black seas included in the Atlantic Ocean.
†Excludes Antarctica and continental areas with no exterior drainage.

Around the world's major land masses, continental margins slope gently into the ocean for an average of 70 kilometers (45 miles) before dropping sharply toward the deep-ocean basins. Above these shallow margins are warm, sunlit coastal oceans, or marginal seas bounded by volcanic islands. The Pacific Ocean, with relatively few of these shallow, marginal areas, has the greatest average depth of the three ocean basins— 3,940 meters (12,900 feet). It is also the largest ocean basin, having nearly the area of the Atlantic and Indian oceans combined. Furthermore, the Pacific Ocean is bordered by deep trenches, particularly along the Asian and South American coasts. High mountains such as the Andes are associated with the trenches that are adjacent to continents, and arcs of volcanic islands (Japan, Indonesia, the Aleutians) occur behind the trenches that lie farther out to sea.

Relatively few rivers discharge into the Pacific Ocean, because it is generally bordered by high volcanic mountains rather than by gently sloping plains with broad river valleys, which surround the Atlantic Ocean basin.

The Atlantic Ocean is relatively narrow (see Figure 1.1) and extends from the North Pole to Antarctica. It has many shallow marginal seas, such as the Gulf of Mexico, and the Caribbean, Mediterranean, and Arctic seas. Atlantic continental shelves tend to be broad, with many river valleys, including the large Amazon and Congo rivers. Heavy freshwater discharge causes some coastal and marginal seas to have lower salt concentrations than occur over most of the world ocean; the Arctic Sea is an example. However, low-latitude seas, such as the Mediterranean and Caribbean, tend to be saltier than average, due to evaporation in the absence of large rivers. Its many shallow nearshore areas make the Atlantic the shallowest ocean, averaging 3,310 meters (10,850 feet) in depth. It has few volcanic islands or mountainous margins.

The Indian Ocean, lying mostly in the Southern Hemisphere, is the smallest ocean basin. Proximity to the large African and Asian landmasses affects its circulation, because continental monsoon winds change direction seasonally and create distinctive surface current patterns. The large Atlantic and Pacific oceans, on the other hand, are little affected by the continents that border them.

Note in Figure 1.1 that the largest part of the world ocean lies within the tropics and subtropics, that is, between latitudes 40 degrees north and 40 degrees south. Most of the sun's radiation is absorbed in this region, then distributed over the earth by winds and ocean currents.

Shorelines, the boundaries between land and water, have been at their present location for only about 3,000 years. They have changed levels many times throughout the earth's history, as ocean-basin shapes and locations have been altered in response to movements of the earth's crust. During successive ice ages, large amounts of water have been held in glaciers and continental ice caps, making sea level lower than at present. Twenty thousand years ago, the ocean surface was about 120 meters (400 feet) below its present level, so much of the now-submerged continental shelf was then dry land (Figure 1.2). Mammoths wandered there, the Greeks built temples that are now partially underwater, and rivers cut valleys across those ancient coastal plains. Asia and North America were connected by a land bridge across the present Bering Sea, permitting large-scale human migrations between continents. Sea level rose when the ice melted, and the ancient coastal plains and river valleys were flooded, creating the coastline we have today.

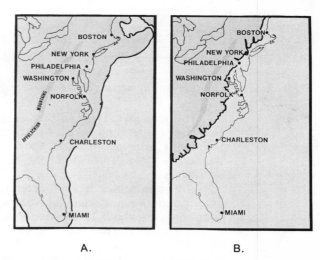

Figure 1.2 *(A) Maximum and (B) minimum sea level on the eastern coast of North America.*

A.   B.

Such important harbors as Boston, New York, Newport, Charleston, and Baltimore were far inland during the ice ages and would be flooded if all present continental ice were to melt during an extended warm period. Present inland river valleys would become bays, and natural harbors would occur where rivers cut through today's Appalachian foothills.

The Ocean Floor

There are three major regions on the earth's surface: continents, ocean basins, and mid-oceanic ridges or rises (Figure 1.3). The continents, together with their shallow, submerged margins, cover about 44 percent of the earth's surface area. Deep-ocean basins separate the continental blocks and account for about 33 percent of the earth. Ocean basins are themselves divided by broad, raised regions that often exhibit a more rugged relief than anything on the continents, known as mid-oceanic ridges or rises (Figure 1.3). These cover about 23 percent of the earth and prevent deep-ocean waters from circulating between ocean basins. Note that mid-oceanic ridges pass through the center of the Atlantic and Indian ocean basins, but that in the Pacific, the East Pacific Rise lies close to the South American coast. Deep trenches occur in ocean basins around the Pacific margin, in the Caribbean Sea, and in the eastern Indian Ocean.

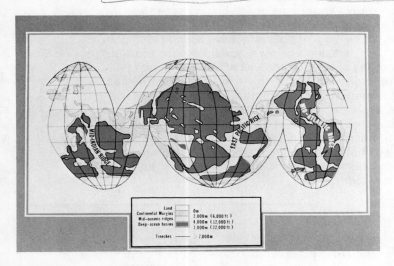

Figure 1.3   *Topography of the ocean floor.*

The outer layer of the earth, that is, its crust, is a relatively lightweight, thin layer comprising only 0.4 percent of the planet's mass. Beneath it is the mantle, which consists of denser material and makes up 68 percent of the earth's mass. The core of the earth lies at the center and is the most dense of all.

The earth's crust is composed of two distinctly different materials—granitic continental crust and basaltic oceanic crust (Figure 1.4). The continents are about 30–35 kilometers (20 miles) thick and have a density of about 2.8 grams per cubic centimeter. (For comparison, ordinary beach sand grains have a density of about 2.6 grams per cubic centimeter.) Both kinds of crust "float" on the underlying mantle, but because the continents are lighter, they project above the more dense (3.0 grams per cubic centimeter), but thinner basaltic crust that underlies the ocean basins. Crustal material is rigid, but the mantle below it is apparently nearly molten and capable of slow movement, rather like hot tar or asphalt.

On stable continental margins where there is no mountain building, for example around most of the Atlantic basin, thick sediment deposits cover the transition between continent and ocean basin. These sediments have been eroded from the land.

Stable continental margins slope gently seaward for tens to hundreds of kilometers offshore and are usually less than 200 meters (600 feet) deep. There is a sharp break in slope at the edge of the shelf, and the continental slope, which is actually the side of the continental block, drops abruptly to a depth of about 2 kilometers (over a mile). Thick sediment deposits, known as the continental rise, extend seaward. These deposits are often several kilometers thick at the base of the slope, gradually becoming thinner with distance from the continent. The continental rise grades into deep-ocean basin at a depth of about 4 kilometers (2.5 miles).

Figure 1.4 *Continental and oceanic crust.*

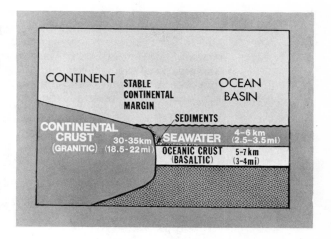

An unstable continental margin is often marked by a deep, narrow trench with associated volcanic mountain ranges on land, or chains of volcanic islands in an arc at sea (Figure 1.5). Marginal ocean basins, partially filled with sediment from the land, often separate trench and island-arc systems from the mainland. These deposits are potential sources of petroleum and natural gas. The Sea of Japan is an example of such a basin.

Figure 1.5 *Structure of unstable continental margins.*

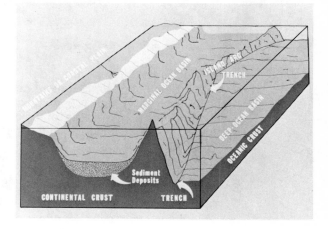

Earthquakes and volcanic activity characterize unstable continental margins, as in Japan. A deep (to 11 kilometers, or 7 miles) trench, long and steep sided, typically separates the deep-ocean basin from a volcanic island arc.

Mid-oceanic ridges are broad, often rugged systems of volcanic mountain ranges that form an interconnected system in all ocean basins (Figure 1.3). They rise from the basin floor to within 2 kilometers (over a mile) of the ocean surface. Their tallest peaks emerge from the sea as volcanic islands. Iceland and the Azores are examples.

A deep, steep-sided rift valley often marks the crest of a mid-oceanic ridge, especially where the ridge is as rugged as the section of Mid-Atlantic Ridge pictured in Figure 1.6. The crest is not a continuous feature but is offset at intervals by long, sharp breaks, or faults, in the sea floor. These faults occur when sections of the ridge under tension suddenly move past each other. Earthquakes and volcanoes are common on the mid-oceanic ridge system, though not all of the system is as rugged as the Mid-Atlantic Ridge. Some sections, such as the East Pacific Rise that passes just west of South America (Figure 1.3), have more gentle relief. Some parts of the mid-oceanic ridge system have ridges of volcanic rock that occur at right angles to the main part of the ridge, as shown in Figure 1.3. The island of Iceland is located on one such ridge, in the North Atlantic Ocean.

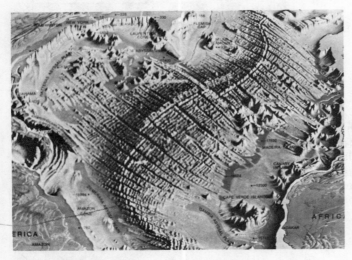

Figure 1.6   *Structure of a mid-oceanic ridge.*

On either side of the ridge, ocean basins average 4–6 kilometers (2.5–3.5 miles) deep. Sediment deposits 200–600 meters (600–1800 feet) deep cover the rough topography of ocean-basin floors. These deposits are thickest at the bases of continents, becoming thinner toward the ridges; the tops of mid-oceanic ridges have virtually no sediment.

Sea-floor Spreading

The earth's crust is made up of rigid plates separated by three kinds of boundaries—mid-oceanic ridges (or rises), deep-ocean trenches, and faults. These crustal blocks are slowly but constantly moving with respect to each other, transported by movements in the nonrigid, deformable mantle below. There are at least six major blocks, and several smaller ones. Each usually contains both continental and oceanic crust. Blocks move away from mid-oceanic ridges, where new crust is formed, and toward the deep-ocean trenches, where oceanic crust is drawn down and resorbed into the mantle. These movements average 2.5–5 centimeters (1–2 inches) per year (Figure 1.7).

Faults are breaks in the earth's crust. Earthquakes occur where tension between blocks builds up, then is abruptly released as rocks break and permit sections of crust to slide past each other. Where blocks containing sections of continents collide, mountain ranges are built as a result of the compression. An example is the Turkey-Arabia-India

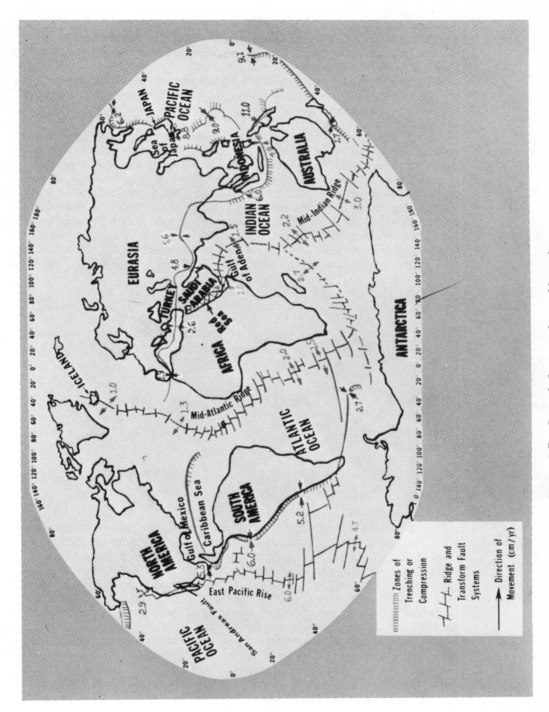

Figure 1.7 *Crustal plates on the earth's surface.*

region, where Africa and the Indian subcontinent are colliding with the Eurasian mainland.

The shape of the Atlantic Ocean basin offers visible evidence of this crustal movement, which is known as sea-floor spreading. The relatively young Atlantic basin began to form about 200 million years ago, when an ancient continent split apart. Figure 1.7 shows the obvious fit of the American continents against Europe and Africa. The Pacific basin is much more ancient, having been formed by sea-floor movements hundreds of millions of years earlier.

Earthquakes and volcanic activity mark the boundaries between crustal blocks. An oceanic crust is dragged or pushed into the mantle beneath a continental block at a trench, earthquakes are generated hundreds of kilometers below the continental margin (Figure 1.8). Only trench and island-arc areas experience such deep quakes. Volcanoes form when sediment deposits and rocks are melted by the high temperatures in the mantle and erupt at the earth's surface. Through long, continued volcanic eruptions, individual volcanoes grow together, forming the typically arc-shaped island groups that occur around northern and western Pacific Ocean margins. Japan, Indonesia, and the Aleutian Islands are examples.

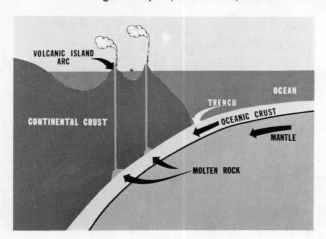

Figure 1.8 *Trench and island-arc system.*

Oceanic crust forms at mid-oceanic rises. Molten rock from the mantle is transported to the surface, and some erupts through volcanoes. As lava, it spreads over the sea floor. Sections of the ridge may be forced apart by this extruded material, and by the molten rock that rises through faults in the volcanic rock. Motion of the underlying mantle away from centers of volcanic activity causes the oceanic crust on either side of the ridge to move away from the ridge. Continental material is propelled in the same direction as the moving sea floor in each crustal block.

Where an area of crustal spreading lies beneath a continental block, the continent may be rifted, or pulled apart, creating a long, narrow gulf. Figure 1.7 shows several such areas. One familiar one is the San Andreas fault-Gulf of California region of the American Southwest coast. Here, Baja California is gradually being drawn away from the mainland, and an associated fault system extends in a northwesterly direction through California, causing frequent earthquakes. Another clearly visible example of continental rifting is the Red Sea-Gulf of Aden region of North Africa, where Saudi

Arabia has split from the mainland and is evidently moving in a northeast direction. New ocean basins eventually form in regions of rifting and formation of oceanic crust.

Continents consist of the relatively lighter components of extruded mantle material. These rocks, largely granitic, have separated from the heavier basaltic rocks of the oceanic crust over billions of years. Aggregated, and prevented by their buoyancy from being drawn down into trenches, continental materials remain at the earth's surface rather than being periodically reworked into the mantle as is oceanic crust. For this reason, some continental rocks may be over 3 billion years old, whereas most of the ocean floor is less than 200 million years old.

Transform faults connect offset sections of a mid-oceanic ridge or rise, where plates slide past each other in opposite directions. Shallow earthquakes occur in these segments. On either side of a ridge or rise, the inactive trace of the fault is marked by a region of mountainous topography and inactive volcanoes known as a fracture zone.

Volcanism and Reef Formation

An active submarine volcano grows into a mountain or an emergent island by frequent eruptions of rock and ash (Figure 1.9). When a volcano is no longer active, waves erode it by cutting into its sides. Further erosion by wind and rain wear down the mountain. Furthermore, the weight of the mountain causes the sea floor under it to sink, because the mantle below is deformed by the extra weight. As a result of these processes, volcanic islands eventually disappear below sea level.

Figure 1.9 *Tenerife, Canary Islands.*

We have learned that volcanoes occur near trenches and along midoceanic ridges in association with generation and destruction of oceanic crust. However, volcanoes that do not appear to be directly associated with sea-floor spreading also occur in many ocean basins, particularly in the Pacific. These volcanoes are formed in places where oceanic crust is weak or cracked and thus penetrable by hot gases and lava from the mantle. Volcanism may also take place above sites of upward movement within the molten mantle. At such locations, sometimes referred to as "hot spots," molten rock apparently rises from deep within the mantle and bursts through the overlying crust. Sometimes, such a spot remains active for many millions of years.

Because oceanic crust moves steadily across such a long-term volcanically active region, a volcano is first formed above the site, then carried past it on the moving sea floor. Chains of volcanic islands, such as the Hawaiian islands, may remain at the surface to record the direction of sea-floor movement during their sequential formation. The islands that have been carried farthest from the volcanic site are the oldest; they also tend to be smaller and less rugged. Active volcanoes occur on the most recently formed island, as is the case for the island of Hawaii (Figure 1.10).

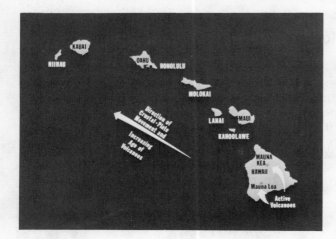

Figure 1.10 *The Hawaiian chain of volcanic islands.*

Submerged sides of volcanoes in tropical regions are sites of coral-reef formation. Corals are tiny colonial animals that live attached to a firm surface. They secrete a cuplike shell of calcium carbonate around their bodies, which, together with the shells of other carbonate-secreting plants and animals, is cemented into a reef structure. Reef-building coral animals live in a mutually beneficial, or symbiotic, association with one-celled algae called zooxanthellae. Because these plants require sunlight, coral reefs grow only in shallow, sunlit waters. Hundreds of kinds of tropical marine plants and animals—walking, swimming, and attached—live on or near the reef surface. Many others live in holes and crevices in and around it.

We have mentioned that volcanic islands eventually subside, because the sea floor below is depressed by the weight of the mountains. Coral reefs attached to the sides of the islands sink with them, into deeper water, where less sunlight penetrates. To remain close to sea level, corals build an ever-higher structure, only the surface of which is inhabited by living organisms.

At first, when a coral reef forms around a volcanic island, it is a near-shore structure known as a fringing reef (Figure 1.11). As the surface of the island is eroded, and also as it sinks, the portion above sea level becomes smaller. However, the submerged reef continues to grow, becoming thicker and higher over millions of years. A barrier reef results, with a shallow lagoon separating the island from the seaward margin of the reef. Debris from plant and animal shells becomes incorporated in the growing reef structure. Eventually, the volcanic island sinks completely below sea level. Corals and other forms of marine life completely cover it with secreted skeletal materials, as well as with sand and other forms of debris. The resulting reef-and-lagoon complex is known as an atoll.

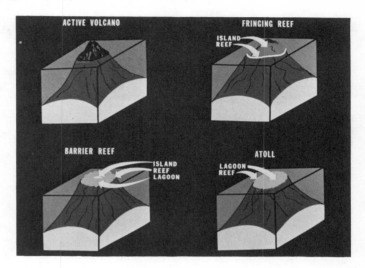

Figure 1.11 *Formation of a coral reef.*

Atolls develop in a variety of shapes and sizes. Sometimes, small islands of sand and detritus form on the outer rim of the reef. In addition, smaller structures of various shapes, called patch reefs, develop in the lagoon. Variety is also a function of the many kinds of coral, each having its characteristic color and configuration.

Midway Island, Bikini atoll, and many other small, typical Pacific islands are atolls formed by subsidence of volcanic islands. Some have shallow lagoons that at a distance appear to be lighter colored than the open ocean. Other atolls are simply a ring below the water surface, with deep, dark blue lagoons. Waves break against the massive fore-reef, which is the seaward side of the structure. Their force is thus dissipated, and the lagoon remains calm, protected from heavy open-ocean waves. Shallow lagoons are easily navigable by small boats and often teem with colorful fishes and other marine life.

Ocean Sediments

The ocean floor, from continental shelf to deep-ocean basin, is covered by sediments, except on mountaintops. These sediments have three sources—particles eroded from continents, shells of marine organisms, and materials precipitated from seawater.

Lithogenous sediments (soil and rock fragments) eroded from continents predominate near land, carried to the ocean by rivers (Figure 1.12). Coarser, heavier particles tend to settle out of moving water first, so they are usually deposited in or near river mouths, where the sand-sized grains form beaches. Fine silt particles may be carried out onto the continental shelf, finally settling to the bottom at depths of 50–150 meters (a few hundred feet).

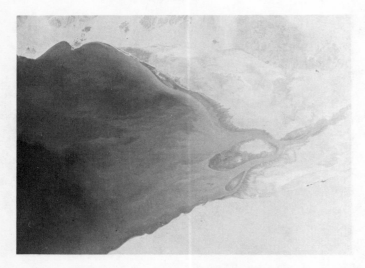

Figure 1.12   *Sediment from the Colorado River, California.*

Sediments accumulate on shallow continental margins at rates of 10 centimeters (a few inches) or more per thousand years. Some are carried down to the continental rise by slumping at the edge of the shelf, or by sudden avalanches of mud and water that flow with great force down channels in the continental slope. Continental-rise deposits may be as much as 11 kilometers (7 miles) deep at the base of a continent, depressing the oceanic crust by their weight. They become thinner in the direction of mid-oceanic ridges, where ocean basins are younger. Sediments derived from land cover preexisting ocean-floor topography, leaving only the mountaintops bare. Some particles from the land, such as desert sands and volcanic ash, are carried over the ocean by winds and eventually settle to the deep-ocean floor.

Biogenous sediments (shells of floating plants and animals) are major contributors to deep-ocean sediment. At tropical and subtropical latitudes, these are largely composed of calcium carbonate. At higher latitudes, silica predominates.

The abundance of these tiny shells is governed by biological productivity of surface ocean waters. In regions where deep waters continuously circulate upward, bringing nitrates and phosphates necessary for plant growth, one-celled algae grow abundantly. They, and the tiny floating animals that feed on them, produce sediment at a rate of about 15 millimeters (0.6 inch) per thousand years. The equatorial region and shallow waters over the continental shelf experience these conditions. However, over most of the open ocean, deep waters move upward very slowly. Therefore, surface plant growth is limited by a lack of phosphates and nitrates. Here, sediment accumulates at a rate of 1 millimeter (0.04 inch) per thousand years.

Biogenous sediments form a blanket over preexisting ocean-bottom topography, blurring sharp outlines in the same way a gentle snowfall does. Below about 5 kilometers (3 miles), however, calcium-based sediments dissolve, leaving only insoluble lithogenous particles.

A third important source of deep-ocean deposits is lithogenous sediment in the form of mineral nodules precipitated from seawater. Over large parts of the deep-ocean bottom, far from land, especially in the Pacific Ocean basin, iron-manganese nodules form very slowly on the ocean floor. These black, potato-sized nodules also contain copper, cobalt, and nickel. They may eventually be mined commercially, replacing scarce supplies on land.

Recent studies of the ages of ocean basins have been based largely on recovery of long sediment cores from the deep ocean. Specially built vessels equipped to drill through a kilometer or more of sediment can penetrate down to basaltic crust. By studying the kinds of fossil shells embedded in the sediments and applying radiometric dating techniques, scientists can learn much about the ages of sediments and the conditions under which they formed. For example, above the most ancient parts of the Atlantic basin, where sediments are many hundreds of meters thick, fossil shells 67 million years old have been recovered. Near the crests of mid-oceanic ridges, in the thin sediments that have recently been deposited on this younger oceanic crust, fossils are no more than a few million years old. Evidence from sediments provided important confirmation of the theory of sea-floor spreading.

Questions

1. What are some ways in which people are affected by the ocean?

2. How is water distributed over the earth's surface? How much of it is in the ocean?

3. What are the three main parts of the world ocean? Which is the Arctic Sea a part of?

4. How much of the earth's surface is covered by water? In which hemisphere is most of the land located? Most of the ocean?

5. Describe some major features of each of the three oceans.

6. What effect does glaciation have on shorelines?

7. What are some characteristics of oceanic crust? Of continental crust?

8. What are some characteristics of stable continental margins?

9. How do unstable continental margins differ from stable ones?

10. Describe a typical mid-oceanic ridge.

11. What is a crustal plate? How is it formed? How destroyed? What kinds of boundaries does a crustal plate have?

12. What determines the direction of crustal-plate motion? What makes a crustal plate move?

13. Explain why earthquakes and volcanoes are associated with deep-ocean trenches.

14. Give some examples of present continental rifting. What causes it?

15. Where do submarine volcanoes form? What happens to a volcano after it forms? How are chains of volcanic islands formed?

16. How does a coral reef originally form? Describe the stages by which it becomes an atoll.

17. What is lithogenous sediment? How does it get into the ocean?

18. What factors govern formation and deposition of biogenous sediment?

19. What are iron-manganese nodules, and how do they form?

20. What information about deep-ocean basins can be gained from deep-sea drilling?

Suggested Readings

Gross, M.G. *Oceanography.* 3rd ed. Columbus, Ohio: Charles E. Merrill Publishing Company, 1976. *Introductory textbook.*

Menard, H.W. *Geology, Resources, and Society.* San Francisco: W.H. Freeman & Company, 1974. *Basic discussion of sea-floor spreading.*

Oceanography: Some Perspectives. San Francisco: W.H. Freeman & Company, 1971. *Collected articles from* Scientific American.

Sullivan, Walter. *Continents in Motion.* New York: McGraw-Hill Book Company, 1975. *Readable treatment of sea-floor spreading.*

MODULE

The Open Ocean

INTRODUCTION

We know that the ocean is very old, and that all its basins are interconnected. Currents carry water, salts, and heat energy to every part of it, and seawater is well stirred by this constant mixing. The relative proportions of its salt constituents are nearly the same everywhere in the world ocean. And because deep-ocean conditions and currents are roughly constant from one year to the next, we say that the ocean is essentially in a steady state. Studies made over long periods of time can be combined to give an accurate picture of ongoing ocean processes.

Changes, both short term and long term, do occur in the ocean. Short-term changes mainly occur in surface or shallow waters, primarily in response to yearly changes in wind direction, rainfall, and sunshine. These changes are often cyclic, and experienced mainly on a local basis. They are likely to affect the coastal ocean, which we will study in the last module of this program.

Long-term changes include alteration of ocean-basin shapes by sea-floor spreading over millions of years. They also include changes in world-wide heating and cooling that lead to expansion or withdrawal of polar ice caps. The very slow addition of dissolved salts leached from continental rock is another example. Long-term changes, occurring very slowly, are not detected in observations taken only a few years apart. Many of the problems that interest oceanographers today involve trying to predict oceanic processes that may occur in the future.

In the first portion of our oceanography program, we studied the structure of ocean basins. In this module, we consider open-ocean processes, and we will see that the shape of those basins controls the flow of ocean currents.

However, that will be only part of the story. We also want to know what drives the currents, which move millions of billions of tons of seawater through the ocean basins. For answers to this question, we first consider some of the physical properties of water itself. Then, we examine some of the effects that adding salt to the water has on those properties. The property of seawater *density* will be particularly important, because density differences control major circulation patterns of the world ocean.

The density of seawater is not constant. It is least at the ocean's surface, greatest at the bottoms of ocean basins. Seawater density also varies horizontally, being greatest at the poles, and least near coasts and in surface waters at the centers of ocean basins. What factors control this distribution?

We know that seawater's chemical composition is uniform. However, seawater varies in the amount of water it contains, or, in other words, in its salt concentration. Increased saltiness, or *salinity,* increases seawater's density. Temperature also plays a major role in determining the density of seawater. So, to understand why density varies throughout the world ocean, and how density differences drive currents, we examine the factors that control salinity variations, in addition to the factors that control the distribution of heat in the ocean.

It may seem that we have introduced a paradox. On one hand, we say that the ocean is well mixed, and in a steady state. Then, we say that differences in important properties occur, and that these differences keep the ocean in constant motion. The resolution of the paradox is found when we construct a budget for heat and water in the ocean.

In one important process—solar heating—the ocean is not homogeneous. Net heating of the earth occurs in the tropics between latitudes 23 degrees north and 23 degrees south. However, we know that the earth does not get hotter year by year. Heat is transported toward the poles, where more energy is lost through radiation back to space than is gained from sunlight. So, for heat on the entire earth, a steady state does exist, but it involves constant motion of winds and waters, which carry heat away from the equator toward higher latitudes.

Water circulation is thus related to heat transport. The amount of water on the earth does not change, although relatively small amounts are removed from the mantle and added to the ocean by volcanic eruptions. Although, as we have learned, the amount of water in the oceans is essentially constant, except for the effects of glaciation and melting, large amounts do move rapidly in and out of the ocean by means of evaporation and rainfall. Subtropical latitudes around 30 degrees north and south are regions of net water loss, and subpolar latitudes from about 40 degrees to about 60 degrees north and south receive more water as rainfall than they lose by evaporation. The water that moves toward higher latitudes in this way carries heat and plays a major role in its global transport.

So, the answer to our question about why seawater is not uniformly dense lies in the unequal heating of the earth's surface by the sun; cold water is denser than warmer water. And when we have studied heat distribution, we will be ready to consider the ocean currents themselves. First, we will study surface currents, which involve a relatively small amount of the oceans, but which affect us directly because they influence our climate. Then, we will conclude our study of open-ocean processes with a discussion of the density-driven deep-ocean circulation, which moves very slowly but involves all the water in the ocean. It is this circulation contained by the deep-ocean basins whose structure we studied in Module I.

By the end of this module, we hope that you will begin to see the world ocean as a dynamic whole, slowly but steadily interrelated by the flow of its currents. In spite of the ocean's vast size, anything that happens in one part eventually affects all other parts. In this study of large-scale open-ocean processes, we will see that the time periods involved are very long, though somewhat shorter than the slow processes studied in Module I.

1. To become familiar with the major properties of seawater that govern its behavior in ocean basins. Specifically, to demonstrate some understanding of heat capacity, latent heat of evaporation, temperature of initial freezing, salinity, and density

OBJECTIVES

2. To understand what a water mass is and how density governs its movements relative to water masses of different densities

3. To know that solar energy is distributed unequally over the earth, and which areas have net uptake or loss of heat

4. To learn some of the factors that determine temperature and salinity distributions in surface and deep waters

5. To be able to explain the concept of global heat transport and know how water is involved in it

6. To gain a general idea why surface currents move in gyres, and to understand what influences the Coriolis effect, Ekman transport, and density-determined sea-surface topography have on the direction of geostrophic currents

7. To understand one specific surface-current process—the formation of eddies at Gulf Stream boundaries

8. To understand the principal mechanism of density-driven deep-ocean circulation and how surface-water properties and ocean-basin shape control deep-ocean circulation

KEY TERMS

heat capacity
latent heat of evaporation
salinity
temperature of initial
  freezing
density
water mass
surface zone
pycnocline zone
deep zone
thermocline
halocline
gyre
western-boundary current

eastern-boundary current
eddy
meander
Coriolis effect
Ekman spiral
Ekman transport
upwelling
geostrophic current
sea-surface topography
Antarctic Bottom Water
North Atlantic Deep
  Water
Antarctic Intermediate
  Water

Now you are ready to begin the audiovisual portion of this module. Select the MEDIAPAK 2 component(s) and proceed. Following is a topical outline of the audiovisual sequence. You will find this outline helpful for reference and review. After completing MEDIAPAK 2, return to this book to perform the exercises.

OUTLINE

Water's Function in Heat Transport Determined by Its Chemical and Physical Properties

Heat capacity of water

Water's heat capacity relative to other substances

Absorption and release of heat by large bodies of water

Latent-heat properties of water

Absorption of heat by evaporation

Release of heat through condensation

Atmospheric gases (e.g., oxygen) dissolved by water

Salinity of seawater

Lowering of temperature of initial freezing by increased salinity

Density of seawater

Density as a function of temperature and salinity

Effect of density on the position of a water mass in a basin.

Distribution of Surface-water Properties—Temperature and Salinity

Effect of latitude on amount of incoming solar radiation

Global heat transport in the ocean and atmosphere

Effect of latitude on surface temperatures

Effect of evaporation and precipitation on surface salinities

Vertical Stratification of the Ocean

Surface zone

Pycnocline zone

Deep zone

Wind-driven Surface Circulation

Movement of surface water in gyres

North and South Equatorial currents and trade winds

Boundary currents and their characteristics

West Wind Drift and westerly winds

The Gulf Stream

Characteristics of the Gulf Stream

Meanders, eddy formation, and paths of eddies

Deflection of wind-driven currents

Coriolis effect

Ekman spiral and Ekman transport

Upwelling and sinking due to the Coriolis effect and Ekman transport

Piling-up of low-salinity surface waters

Deep-ocean, Density-driven Circulation

Formation of polar surface-water masses

Sinking and north-south movement of dense waters

Antarctic Bottom Water

North Atlantic Deep Water

Antarctic Intermediate Water

Exercises

1. The number of calories required to raise the temperature of 1 gram of a substance by 1°C is termed its
   A. latent heat.
   B. density.
   C. thermocline.
   D. heat capacity.

2. Adding salt to water lowers the water's
   A. density.
   B. heat capacity.
   C. temperature of initial freezing.
   D. salinity.

3. Density varies most markedly with depth in the
   A. surface zone.
   B. pycnocline zone.
   C. deep zone.
   D. upwelling zone.

4. Water that has been away from the surface for a long time may contain little or no
   A. sea salt.
   B. dissolved oxygen.
   C. nitrogen gas.
   D. carbon dioxide.

5. The salinity of seawater is a measure of how much _____ is dissolved in it.

6. A less-dense water layer is in a stable position when it is _____ a more dense water layer.

7. A large, nearly closed current system is called a _____.

8. *True or false:*
   A. Very dense water is likely to have high temperature and low salinity.
   B. Subtropical regions are characterized by high precipitation and low evaporation.
   C. Water movements help transport heat from regions of net heat gain to regions of net heat loss.

34

    D.   Wind blowing across deep water causes the water in the surface zone to move 45 degrees to the right of the wind in the Northern Hemisphere.

9.   Match the following:

    A.   The Gulf Stream is a(n)

    B.   Water samples can be taken with a(n)

    C.   Surface waters in an area of excess evaporation usually have

    D.   A region of marked temperature change is called a(n)

    E.   A major current gyre includes one relatively weak, shallow

    F.   An eddy forms from a(n)

    G.   The path of a moving object traveling a long distance through the ocean is apparently deflected by the

    H.   Each water layer sets in motion the one below, forming a(n)

    I.   Geostrophic currents are controlled by the Coriolis effect and

    J.   When surface waters are driven away from a coast, the resulting vertical water movement is termed

    K.   Ekman transport creates "hills" of

    L.   Deep-ocean currents are driven primarily by

    M.   The densest water mass in the ocean is the

    N.   Knowing the distribution of sea-surface topography permits oceanographers to predict the direction of a(n)

1.   low-density water
2.   upwelling
3.   thermocline
4.   geostrophic current
5.   Ekman spiral
6.   meander
7.   Coriolis effect
8.   density differences
9.   gravitational attraction
10.   western-boundary current
11.   eastern-boundary current
12.   Antarctic Bottom Water
13.   Nansen bottle
14.   high salinity

SUMMARY

Temperature, Salinity, and Density

Water plays an important role in the transport of heat over the globe, and in moderating the earth's climate. One reason for this is its high heat capacity, meaning that relative to other substances water can absorb a great deal of heat while experiencing little rise in temperature. We observe this during a summer day at the beach, for example. In the morning, the sand is cool, having lost yesterday's heat by radiating it back to space during the night. However, as the day wears on, the sand may become unpleasantly hot. The water, on the other hand, varies perhaps 1 or 2 degrees in temperature over a 24-hour period.

Large bodies of water moderate the climate of nearby land by absorbing heat from the atmosphere in summer and releasing heat in winter. Near coasts, relatively warm, damp winters and pleasant, cool summers are likely to be the rule. At the same latitude, but farther inland, summers may be extremely hot while winter temperatures fall below -18°C (0°F).

Water not only absorbs a great deal of heat; it can transfer that heat from place to place on the globe. Five hundred eighty-five calories of heat are required to evaporate 1 gram of water at a temperature of 20°C (68°F) (Figure 2.1). (Note that the calorie used here is only 1/1,000 the calorie commonly used to calculate food energy.) This amount of heat, called the latent heat of evaporation, is stored in water vapor and released to the environment when water condenses as rain, snow, or dew. Thus, if water evaporates in a warm region and condenses in a cool one, heat is transferred from the warmer to the cooler area.

Another important property of water is its ability to dissolve gases from the atmosphere. Nitrogen and oxygen, which make up 99 percent of atmospheric gases, are also the most abundant in surface seawater.

Marine animals remove oxygen from ocean waters in their metabolic processes, giving off carbon dioxide in the process, just as land animals do. Water that has been away from the ocean surface for a long time may contain almost no dissolved oxygen. Oceanographers measure this property and use it to trace water masses that sink below the surface and move for hundreds or thousands of years through deep-ocean basins.

Carbon dioxide (CO_2) dissolves readily in seawater. It is used by plants and some animals to make carbonate ($CaCO_3$) shells, which ultimately form the carbonate sediments that cover large areas of the ocean floor.

Salinity is a measure of salt content. Most of the water in the oceans has a salinity of about 35 parts per thousand (35‰), which means that

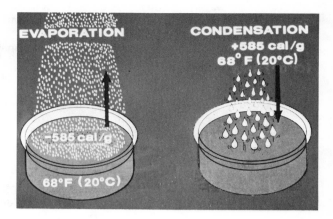

Figure 2.1 *Latent heat and energy transfer.*

each 1,000 grams of seawater contains 35 grams of dissolved salts. The composition of these salts, and the relative proportions in which they occur, are constant throughout the world ocean (Figure 2.2). Sodium, chloride, and sulfate together make up more than 93 percent of sea salt. Another 6 percent consists of magnesium, calcium, and potassium. Sixty-six different elements are also present but make up less than 1 percent of the total.

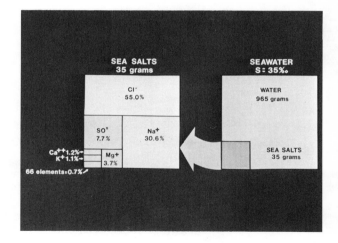

Figure 2.2 *Composition of sea salt.*

Since the major elements occur everywhere in an essentially constant ratio, oceanographers measure salinity by determining the amount of a major element (usually chloride) in a seawater sample. Another way is to measure the conductivity of seawater, since conductivity and salinity are directly proportional.

Two properties of seawater controlled by salinity are important in ocean processes. These are the water's temperature of initial freezing and its density. Increasing salinity causes seawater to begin freezing at lower temperatures. However, when seawater freezes, the ice is made of pure water. Sea salt is excluded from the ice structure and remains in unfrozen water, thereby making the water more salty. If water temperatures continue to drop, more sea ice forms, and salinity of the remaining brine continues to increase.

The property that controls ocean currents and the vertical position of a water mass in an ocean basin is the seawater's density, that is, its weight per unit volume. Let us demonstrate density by preparing two dishes, each containing 100 grams of fresh water at room temperature. If we add some salt to the water in the first dish, we make the water heavier without significantly increasing its volume. We have thus increased its density. Now, if we cool that same dish of water by refrigerating it for a few hours, we cause the water to shrink very slightly without becoming any heavier. Again, we have increased the density of the water.

Suppose we now add blue dye to the cold, salty water and red color to the dish of fresh water at room temperature. If we gently allow the less dense red water to flow over the surface of the more dense blue water, the red water will remain there as a discrete layer, without mixing, because a less dense liquid will float on a more dense one. If we take a third measure of water whose density is between the red and the blue and allow drops of it to fall on the surface of the water in the dish, the drops will pass through the fresh water and form a layer between it and the denser water below, as shown in Figure 2.3. The layers will not mix unless we stir them, and the greater we make the density difference between the water masses, the more energy will be needed to mix them.

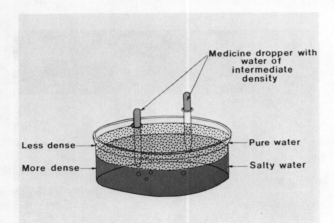

Figure 2.3 *A stable density-stratified system.*

Seawater in ocean basins behaves like the water in the dish. The least dense seawater floats at the surface, with layers of increasingly dense water below. However, if surface waters become saltier due to freezing of sea ice, or if they become chilled by being carried to a higher latitude, small parcels of surface water begin to sink through the less dense water layers below. When a parcel reaches the appropriate depth for its density, sinking ceases. Uniformly dense water parcels unite to form a discrete layer, with less dense water above and more dense water (or the ocean floor) below. Figure 2.4 shows how salinity and temperature govern seawater's density.

Temperature and salinity measurements may be made at any depth in the ocean using a sampling bottle to which a precision reversing thermom-

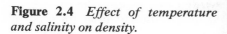

Figure 2.4 *Effect of temperature and salinity on density.*

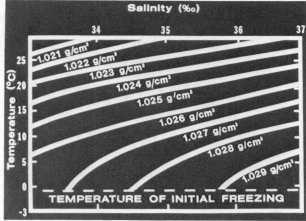

eter is attached. The bottle is lowered on a wire to the desired depth. A trigger mechanism is released, causing the bottle to invert and take a water sample. At the same moment, water temperature is recorded. Water density is calculated from the temperature and from the salinity of the sample, which can be measured in a laboratory aboard ship. The entire device is called a Nansen bottle, in honor of the Norwegian explorer and oceanographer Fridtjof Nansen (1861–1930), who invented it.

To understand how surface-water density is distributed, we look first at the way sunlight is received at the earth's surface (Figure 2.5). Between latitudes 23 degrees north and south, the sun is nearly overhead throughout the year. Here, much of the incoming solar energy is absorbed by the ocean. At higher latitudes, especially near the poles, less energy strikes the earth per unit area to be absorbed by the ocean. Also, much heat is lost during the long winters. As a result of this unequal heat distribution, subpolar and polar areas lose more heat by radiation to space than they gain by direct solar heating, and tropics and subtropics are regions of net heat gain.

Figure 2.5 *Distribution of solar radiation on the earth.*

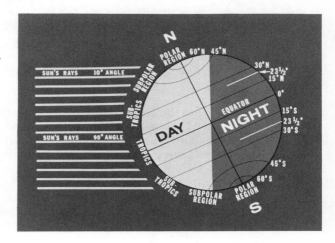

Ocean waters absorb the sun's energy at an average rate of 0.25 calorie per square centimeter. But the equator does not get hotter year by year, nor do the poles become colder. Tropical and subtropical waters are usually around 20°C (68°F) or warmer. Near the poles, surface temperatures remain around -2°C(28.5°F), which is the temperature of initial freezing for seawater of salinity 32‰. (Land temperatures drop much lower.) This temperature regime is maintained by the regular transport of heat from the equator toward the poles. Water in the atmosphere and ocean, with its high heat capacity, is the principal medium of transport.

Surface-ocean temperatures, then, are governed by latitude and tend to run east-west, especially across the broad Indian and Pacific oceans (Figure 2.6). Note that the Atlantic Ocean shows some variation in this climate pattern. West directed surface currents are deflected by the conclimate pattern. East-west directed surface currents are deflected by the continents in both hemispheres. Warm subpolar waters are carried northward along the North American coast, as shown by the upward-trending temperature bands in the North Atlantic, at the right. In the southern Atlantic and Pacific oceans, cool (25°C) waters are carried toward the equator by currents to the west of Africa and South America. These deflections of the east-west distribution of surface temperature are an important feature in the global heat transport system.

Figure 2.6 *Distribution of temperature in the surface ocean.*

Distribution of evaporation and precipitation over the ocean surface is affected by the amount of solar radiation received, and also by wind speed and relative humidity. At subtropical latitudes, winds are warm and steady, causing more evaporation than precipitation. However, along the equator, even though the sun is directly overhead, winds are weak, and regional cloudiness is typical. This results in an abundance of rainfall and less evaporation. Therefore, this is a region of net water surplus. This is also the case at subpolar latitudes. Near the poles, however, little water either enters or leaves the atmosphere; both precipitation and evaporation are very low.

Salinity of surface waters is controlled mainly by discharge of fresh water from rivers, which dilutes coastal waters, and by evaporation from the ocean surface (Figure 2.7). Salinity is highest in regions of excess evaporation, for instance at the centers of oceans far from land. Salinities are also high near desert shores, for instance off Saudi Arabia. In the subpolar North Pacific, where rainfall is heavy, salinities are low.

Salinity (‰)

Figure 2.7 *Distribution of surface-water salinity.*

To summarize what we have learned about temperature and salinity distribution in open-ocean areas, surface salinity varies by only a few parts per thousand (35‰ is the average salinity). Most of the ocean is in the 33‰–37‰ range, although salinity may be much less near the mouth of a large river. Surface-ocean temperatures, on the other hand (Figure 2.6), may vary 25°C or more between the equator and the poles.

It is important to realize that temperature and salinity changes, as well as uptake or release of atmospheric gases, can occur only at the ocean's surface. This means that all water masses acquire their characteristic temperatures and salinities while at the surface. However, most of the ocean, as we see in Figure 2.8, is not in contact with the atmosphere, and it lies too far below the surface to be heated by the sun, or cooled by radiation of heat back to space or by evaporation. Below about 2 kilometers (1.2 miles) depth, it is cold, generally about 3.5°C (38°F), although clearly above the freezing point. (Note that this average is well above the temperature of polar surface-ocean water masses.) How is it possible that processes acting at the surface can affect the vast, seemingly inert mass of subsurface water?

Depth Zones in the Ocean

Before we try to answer this question, let us learn just a little about the ocean's layered structure. Then, we will study surface-water movements in preparation for our discussion of how the entire ocean is interconnected by processes that affect every part of its breadth and depth.

There are three major depth zones in the open ocean (Figure 2.8). The surface zone, about 100 meters (300 feet) deep, receives heat from the sun, and fresh water from rain and river discharge. At certain latitudes, it also experiences excess evaporation. Wind waves and associated water turbulence constantly mix surface-derived heat and water throughout the surface zone, making it vertically homogeneous in temperature and salinity.

The floor of the surface zone is the depth at which surface mixing processes are no longer effective. Here begins the pycnocline zone, a layer in which density changes markedly with depth. At the top of the pycnocline, waters are only slightly denser than in the surface zone above. Temperature drops rapidly with depth in this zone, creating a thermocline, or steep temperature gradient. Salinity may also increase with depth, forming a marked salinity gradient, or halocline. This sudden drop in temperature and increase in salinity give pycnocline-zone waters great stability, so great energy is required to mix them. It is this stability which prevents pycnocline waters from mixing with surface-zone waters above or with the deep zone below.

Figure 2.8 *Depth zones in the open ocean.*

The deep zone usually begins at about 2 kilometers (1.2 miles) depth. Note in Figure 2.8 that this region below the pycnocline zone contains about 80 percent of ocean waters. Its temperatures average around 3.5°C (38°F), its salinities around 35‰. Since these figures are substantially the same as those for the ocean as a whole, it is apparent that surface-zone properties prevail throughout only a very small part of the world ocean. Note also that deep-zone waters extend to the surface at high latitudes where no pycnocline exists (Figure 2.8). Near the poles, waters are cold and dense throughout the water column, so surface-water conditions are essentially the same as those at great depth.

Ocean Currents

The pycnocline acts as a floor under wind-driven surface-ocean currents. Prevailing winds drive surface waters in generally east-west directions, but these currents are deflected by continental blocks and so move in the closed loops, or gyres, as shown in Figure 2.9. Sun-warmed waters from the North and South Equatorial currents are deflected poleward at the western margins of Atlantic and Pacific Ocean basins. Moving away from the equator, they carry warm water to higher latitudes. Here, these waters are cooled before being returned to the tropics at the eastern boundaries of ocean basins. Note that around Antarctica the West Wind Drift flows eastward without interruption.

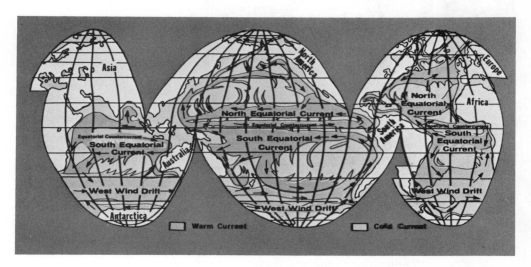

Figure 2.9 *Surface-ocean current systems.*

Each current gyre consists of two currents directed east and west, and two directed north-south. North and South Equatorial currents flow westward, driven by trade winds blowing from the east. Part of their flow is returned in an Equatorial Countercurrent, which moves eastward in a belt of weak and variable winds called the doldrums. However, most Equatorial Current water is deflected at the western side of the basin and flows toward the poles as fast-moving western-boundary currents. These deep, narrow currents travel seaward of the continental slope at speeds of 35–110 kilometers (20–70 miles) or more per day. They carry warm water from the tropics to the high-latitude margins of ocean basins, where water is deflected westward by continental margins and westerly winds. Here, these waters are cooled and carried toward eastern ocean-basin margins, then transported toward the equator in broad, slow-moving, shallow eastern-boundary currents. During this part of the current cycle, water speeds seldom exceed a few kilometers (2–4 miles) per day.

In the Northern Hemisphere, western-boundary currents are the Kuroshio in the Pacific and the Gulf Stream in the western Atlantic Ocean. The Gulf Stream has been studied extensively, since the warm temperatures of its surface water makes it an easy current to trace. Temperature-sensitive cameras mounted on earth satellites give us a picture of the Stream and its surrounding waters (Figure 2.10). This swift current, 100–200 meters (300–600 feet) deep, flows northward beyond the continental shelf and slope. It is several degrees warmer than the Sargasso Sea region of the North Atlantic to the southeast, and 10°C or more warmer than slope waters to the north. Continental-shelf waters were the coldest of all when these data were taken, in late April 1974.

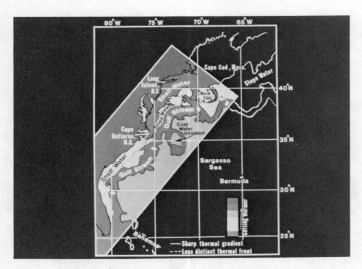

Figure 2.10 *Meanders and eddy formation in the Gulf
Stream.*

Meanders in the Gulf Stream, both beyond its North Wall on the continental slope, and southward in the Sargasso Sea, form loops of warm
water that may break free of the Stream itself. In some cases, waters from
the continental slope may intrude in a meander that breaks off in the Sargasso Sea. This water moves southwestward as a cool eddy. Conversely,
Sargasso Sea water may be enclosed in a meander that eventually forms a
warm eddy which moves southwestward along the continental slope. Such
a warm eddy is particularly easy to trace, since its low-density waters float
on the top of the cooler slope waters around and beneath it.

Water temperature and salinity data taken by merchant vessels, fishing boats, and Coast Guard and Navy ships are used in following the
course of eddies. These data are transmitted to oceanographic facilities,
and maps are drawn to show the path of an eddy from the time it forms
until it is resorbed farther to the south. This resorption often occurs where
the Gulf Stream comes against the continental shelf, off Cape Hatteras
and along the Atlantic coast of Florida.

The Coriolis Effect

We have established that long-distance surface currents are driven east
and west across ocean basins by prevailing winds. In fact, this is an over-
simplification of the relationship between winds and surface-ocean circulation. A complication arises because the earth's eastward rotation causes
the thin layer of water closest to the surface to appear to be deflected 45
degrees to the right of the wind (clockwise) in the Northern Hemisphere,
and to the left of the wind (counterclockwise) in the Southern Hemisphere.
This apparent deflection of a moving object with respect to the earth's surface is known as the Coriolis effect, in honor of the French scientist who
first explained it.

To understand the Coriolis effect, picture a water parcel moving from the equator toward the North Pole on the globe pictured in Figure 2.11. At the equator, the water parcel moves eastward with the rotating earth, at about 1,600 kilometers (1,000 miles) per hour. Traveling northward, it maintains the same net eastward motion, because it is not actually attached to the earth's surface but floats freely above it. (The path of a car or a train is not affected by the Coriolis effect.) Note that at latitude 30 degrees north, the earth's surface is only moving eastward at 1,394 kilometers (866 miles) per hour, while the current still has its original eastward speed of 1,600 kilometers (1,000 miles) per hour. So, the water's net eastward speed is greater than that of the earth beneath it, and it appears to have been deflected eastward, that is, to the right (clockwise) in the Northern Hemisphere.

Figure 2.11 *The Coriolis effect.*

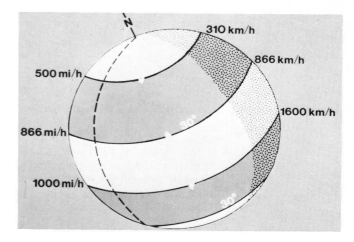

A westward- or northwestward-moving water parcel traveling at a constant speed moves eastward more slowly than the earth, so it appears to be deflected away from the equator. Again, this is to the right in the Northern Hemisphere. In the Southern Hemisphere, westward-moving water is also deflected eastward, away from the equator, but here, eastward is to the left, or counterclockwise.

The amount of apparent deflection depends on the speed and latitude of the moving water parcel, or current. Deflection is zero at the equator and greatest at the poles.

To an observer on the moon, the path of a moving object would not appear to be deflected. That observer would see that the object moves in a straight line while the earth turned eastward beneath it.

The Ekman Spiral

The Coriolis effect plays a major role in controlling wind and current direction. A thin surface-water layer is driven 45 degrees to the right of the wind (in the Northern Hemisphere), and this layer drags at the layer below, setting it in motion. The process continues downward, with each layer being further deflected by the Coriolis effect. Deeper layers move more slowly, because momentum is lost with each transfer between layers.

The downward-spiraling current system set up by this process is called an Ekman spiral (Figure 2.12). At its base, usually at a depth of about 100 meters (300 feet), water moves very slowly in a direction opposite the surface current. This point is usually considered to be the bottom of the wind-driven surface circulation.

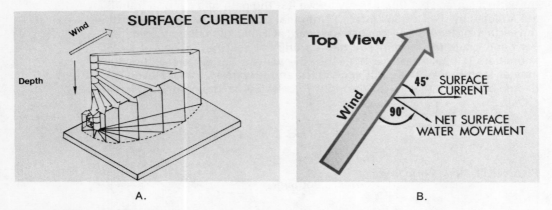

Figure 2.12   *(A) Ekman spiral. (B) Ekman spiral viewed from above.*

Net water movement of the surface layer as a whole is 90 degrees to the right of the wind in the deep ocean. This figure is based on water movement in an idealized, homogeneous ocean, having no pycnocline and no boundaries. However, since our ocean does not fit these conditions, actual wind-induced water movements may be different from predictions based on a theoretical model. For example, the angle between wind direction and surface-water movement varies from 15 degrees in shallow coastal waters to the maximum of 45 degrees in parts of the deep, open ocean. The net movement of waters whose direction is determined by an Ekman spiral is termed Ekman transport.

Vertical water movements, as well as horizontal ones, are caused by the Coriolis effect and associated Ekman transport. Let us see what happens when winds blow parallel to a coast, across the shallow coastal ocean. For example, winds from the north blow southward along the Pacific coast of the United States. Wind-driven surface waters also begin to move southward but are deflected clockwise by the Coriolis effect and Ekman transport. This results in a net seaward motion of the warm, low-density surface-water layer immediately adjacent to the coast. Cold, dense waters then rise to the surface in this area (Figure 2.13). This process, known as upwelling, causes the cool, foggy spring and summer weather characteristic of the coast along Washington, Oregon, and northern California.

Geostrophic Currents

During the fall and winter, winds blowing from the south along the same coast cause surface waters to move toward the coast instead of away from it. Low-density surface water piles up against the edge of the continent and sinks, moving seaward at depth along the continental shelf and slope (Figure 2.13).

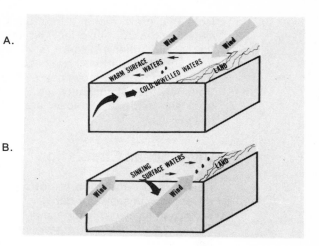

A.

B.

Figure 2.13 (*A*) *Wind-induced upwelling of subsurface waters.* (*B*) *Wind-induced sinking of surface waters.*

The clockwise direction of net water movements in the open oceans of the Northern Hemisphere also causes low-density surface waters to accumulate at the centers of ocean basins, where it piles up in low "hills." This gentle sea-surface topography seldom exceeds 2 meters (6 feet). The "hills" exist because a water column of lower average density occupies more space than a water column of the same mass whose average density is greater. Thus, the surface of a low-density area is elevated relative to an area of average density.

A "hill" of low-density water cannot be measured directly but instead is located by taking precise measurements of seawater temperature and salinity over a wide area and calculating seawater density. The distribution of surface-water densities is calculated and plotted on a map. Areas of low-density waters are identified, and their heights above a standard level are calculated. Surface-current direction and speed in each area are estimated from the calculated topography. The currents tend to flow around the "hills" of water. Therefore, the shape and location of these "hills" indicate current direction. Currents are strongest on steep slopes, so the steepness of sea-surface topography indicates the current speeds.

Figure 2.14 *Geostrophic current resulting from a sloping sea surface and gravitational attraction balanced by the Coriolis effect.*

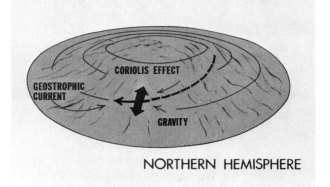

CORIOLIS EFFECT

GEOSTROPHIC CURRENT

GRAVITY

NORTHERN HEMISPHERE

Note that gravitational attraction, directing the current downhill, is balanced by the Coriolis effect, directing the current to the right (in the Northern Hemisphere) (Figure 2.14). Predictions of current direction made from such calculations have been confirmed by current measurements and tracking of iceberg movements. The so-called geostrophic currents, controlled by the balance between Coriolis effects and gravitational attraction from elevated sea surfaces, are the major surface currents in the open ocean.

Deep-ocean Circulation

Deep-ocean circulation is not affected by winds. Rather, it is driven by differences in seawater density. The coldest, most dense water masses in the ocean form at the surface of polar oceans, especially the Antarctic (see Figure 2.8). Freezing of sea ice increases the salinity of these waters, because although seawater freezes at extremely high latitudes, it may not melt there. Sea ice tends to drift toward temperate latitudes, where summer warming eventually causes melting. A cold, dense water mass known as the Antarctic Bottom Water sinks to the bottom of southern oceans, where it spreads out along the bottom of all three deep-ocean basins (Figure 2.15). It moves slowly northward, far into the Northern Hemisphere, while gradually becoming warmer. (Note that the shape of deep-ocean basins restricts this flow to a basically northward direction.) Gradually, over hundreds of years, deep water diffuses upward through the pycnocline and eventually returns to the ocean surface.

Figure 2.15 *Direction of water movements in the deep ocean.*

Arctic water masses flow into the Atlantic Ocean, over the volcanic ridge on either side of Iceland (see Figure 1.3). The so-called North Atlantic Deep Water is identifiable at all depths in the North Atlantic Ocean, as far south as about latitude 35 degrees north. South of that point, it is still recognizable as a distinct water mass by its characteristic temperature, salinity, and dissolved oxygen content. At subtropical and tropical latitudes, North Atlantic Deep Water floats above the denser mass of Antarctic Bottom Water moving northward along the ocean floor.

Another, less dense water mass, the Antarctic Intermediate Water, is carried below the ocean surface at the Antarctic Convergence. This area of sinking waters occurs where the West Wind Drift converges with eastward-moving water from Southern Hemisphere current gyres (see Figure 2.9). Antarctic Intermediate Water forms a layer above the denser Antarctic Bottom Water and is recognizable to about latitude 20 degrees north by its characteristic properties of temperature, salinity, and dissolved oxygen.

We can now return to the question posed earlier in this section: How is it possible that processes acting at the surface can affect the ocean as a whole? A few answers are possible. First, we know that the great water masses of deep oceans are formed in high-latitude surface waters. We have also seen that the temperature and salinity of polar oceans depends in part on the amount of heat received at low latitudes, because it is from that source that heat is transported poleward. The amount of heat and water carried in surface currents and water vapor, together with the speed and direction of that transport, partially controls the properties of polar waters. Therefore, any major worldwide change in salinity or temperature of surface waters must eventually affect the density of deep-ocean water masses and their distribution in ocean basins.

Questions

1. Explain what is meant by the statement "Water has a high heat capacity."

2. How does a large body of water affect adjacent land climate?

3. Discuss two ways water is involved in global heat transport.

4. How does freezing of sea ice affect the salinity of seawater?

5. What changes in water properties can cause sinking of surface waters in ocean basins? Explain your answer.

6. What factors control surface-temperature distribution in the ocean? Name two, and explain the effect of each.

7. What factors control surface-water salinity? Give a general description of how salinity is distributed in the surface ocean.

8. Name the three major depth zones in the ocean, and list the most conspicuous characteristics of each.

9. What is current gyre? How is it set in motion? How are water and heat transported in a gyre?

10. List at least three characteristics of western-boundary currents. Contrast these with the characteristics of eastern-boundary currents.

11. What is a current meander? What is an eddy? Explain how the two are related.

12. What apparent influence does the Coriolis effect have on surface currents in the Northern Hemisphere? In the Southern Hemisphere?

13. How would an observer on the moon view the path of an object moving northward over the earth's Northern Hemisphere? How would that view differ from that of an observer on the earth?

14. As a result of the Coriolis effect, in what direction does surface water in the Northern Hemisphere move with respect to the wind that drives it?

15. Explain how an Ekman spiral forms. What is the net direction of Ekman transport with respect to the wind direction?

16. Explain the horizontal and vertical water movements that occur when a wind blows from the north parallel to the west coast of a Northern Hemisphere continent.

17. What causes "hills" of low-density water to form at the center of ocean basins? How do they affect surface currents? How do oceanographers identify such "hills"?

18. What are the characteristics of water masses that form in polar oceans? What happens to these water masses?

19. How does the shape of deep-ocean basins influence deep-ocean circulation?

20. Explain what is meant by the statement "Anything that affects water properties anywhere in the ocean eventually affects the ocean as a whole."

Suggested Readings

MacIntyre, Ferran. "Why the Sea Is Salt." *Scientific American* 223 (1970): 104–15.

Munk, Walter. "The Circulation of the Oceans." *Scientific American* 193 (1955): 96–102.

Pickard, G.L. *Descriptive Physical Oceanography.* New York: Pergamon Press, 1964. *Intermediate difficulty.*

Stommel, Henry. "The Circulation of the Abyss." *Scientific American* 199 (1958): 85–90.

Vetter, R.C. *Oceanography: The Land Frontier.* Voice of America, Forum Series. Washington, D.C.: U.S. Information Agency, 1974. *Papers on basic aspects of oceanography.*

MODULE 3

Waves, Tides, and Shoreline Processes

INTRODUCTION

In previous sections, we dealt with large-scale processes that affect entire ocean basins and operate over months or years. In this module, we examine small processes, measured in hours, minutes, or even seconds. Paradoxically, it is these minor-scale oceanographic phenomena that most intimately affect the lives of shore-dwelling populations.

Waves can be seen at any ocean beach or along the shore of any large body of water. Ancient peoples living on the shores of the Mediterranean or Aegean seas must have watched the play of water surfaces, just as we do. Some of them no doubt observed that waves approach a shore from great distances, even when no wind is blowing to stir the water surface. Others surely noted that a stick in the water merely bobs up and down in the same place, while waves passing beneath it travel on toward the beach. Some among these ancient wave-watchers may have realized that waves are nothing but moving forms; the water stays in the same place. But it was not until the nineteenth century that mathematicians began to study waves and develop theories to explain their behavior.

Tides were of great interest to ancient seafaring peoples, in part because they affect ship launchings. Aristotle (384–322 B.C.), who was quite interested in oceanography, paid little attention to tides except to note that in narrow straits the sea "swings to and fro" with a motion imperceptible in larger ocean areas. He had no notion of what causes tidal variations, and, in any case, he observed the Mediterranean Sea, where tidal effects are minimal. Aristotle's failure to provide an explanation for tides gave rise to a legend that was repeated in 1675 by Richard Bolland:

> The great Master of Philosophy drowned himself because he could not apprehend the Cause of Tydes; but his Example cannot be so prevalent with all, as to put a Period to other Mens Inquires into this Subject.

Other men did inquire. Pliny the Elder (A.D. 23–79), the great Roman natural historian, knew that the moon causes tides, and that different ocean areas respond in different ways to its passage. Julius Caesar (100–44 B.C.), reporting on a storm that damaged his fleet in Britain, attributed the disaster to a full moon when he stated, "the period of high tides at sea, and the Romans did not know it." The Venerable Bede (A.D. 673–735) even knew that precise tidal predictions must take into account the complete lunar cycle of 19 years.

However, not everyone was convinced of the lunar effect. Caesar's contemporary Pomponius Mela of Spain thought that the earth's breathing might be responsible for tides. Many medieval writers supported the theory of a great whirlpool in the sea that sucks water in, then spews it out again.

During the Renaissance, Sir Isaac Newton (1642–1727) developed an equilibrium theory of tides. His theory correctly explained the influence of the sun and moon on water movements but failed to take into account the individual response of ocean basins to the tide-generating forces.

Until the mid-twentieth century, studies of waves and tides made slow progress. Research on tides was largely confined to preparing tables for prediction of height and timing of tides at various ports. For wave studies, two approaches existed: deep-water waves were treated by mathematicians as problems in theoretical hydrodynamics, while empirical observers described the behavior of real waves as they appeared from shipboard. Attempts to integrate these two approaches failed until, with the development of amphibious warfare during World War II, the problem of waves was transformed from a purely intellectual exercise to a matter of vital concern to the war effort.

Before the war began, a team of California scientists had begun to try to predict wave conditions. Since the size of waves depends on a combination of wind speed, the length of time the wind has been blowing, and the size and shape of the body of water across which it blows, these scientists were able to work out a method for computing wave height from these three known factors. Next, they attacked the problem of how waves behave upon entering shallow water. They found that it is possible to calculate the depth at which a wave will break when nearing a beach on the basis of its original dimensions in deep water. Field manuals explaining how to forecast waves and surf were developed for use in landing troops, for instance on the Normandy beaches in 1944. To improve the accuracy of the forecasts, additional factors were considered, such as the affect of *tidal currents* and irregularities of the four assault beaches.

By spring of 1944, techniques of wave prediction had achieved 80 percent accuracy when based on weather conditions two or three days ahead of a desired target date. Aerial photographs were used to check the accuracy of predictions. The planned June 5 landing date promised favorable tides but was postponed for a day on the basis of a cold front moving through the English Channel and bringing rough seas. On June 6, when the troops landed, they encountered exactly the surf conditions that had been predicted: three- to four-foot waves at the unloading zone ten miles offshore, and four- to six-foot breakers on the beaches. But nearshore currents and choppy seas proved more of a problem that had been anticipated, and the landing operation was considerably hampered.

Later in the war, when landings were being made on one Pacific island and atoll after another, forecast groups continued to help plan the invasions. In that theater of war, however, a lack of knowledge about reefs and other submarine hazards intoduced a dangerous amount of guesswork into landing plans.

After the war, wave and tide research continued along somewhat different lines. Forecasting of storm waves and tidal conditions is important for protecting coastal areas against seismic sea waves and hurricane damage. Even when no unusual weather conditions arise, beaches erode, and waterfront buildings are undermined by wave action and tidal currents. To prevent these occurrences and to protect valuable coastal property, it is necessary to know what wave and tide conditions commonly prevail on a particular stretch of coast. Then, structures such as roads, hotels, and beach cottages should be planned so that their construction does not damage *dunes* or other natural formations that serve as barriers to storm waves that attack the coast. Although great strides have been made in this type of coastal management, more work needs to be done.

1. To know how a sea is developed, how it is transformed into swell, and how the original wind energy is finally dissipated

2. To be able to identify the parts of an ideal progressive wave, and to know how water particles moves in a wave, and how a shallow-water wave differs from a deep-water wave

OBJECTIVES

3. To understand how the attraction of the sun and the moon govern tides, and to learn what factors other than gravitational attraction determine the height and timing of tides

4. To be able to describe the general behavior of a tidal current that moves as a progressive wave, one that moves as a standing wave, and one that moves as a rotary current

5. To know some of the mechanisms by which wave action shapes coastlines

capillary wave	breaker
sea	progressive wave
swell	crest (of a wave)
wave period	trough (of a wave)

KEY TERMS

wave height   mixed tide
wave length   tidal current
shallow-water wave   flood current
surf zone   slack water
standing wave   ebb current
antinode   reversing tidal current
node   rotary tidal current
tidal range   inlet
spring tide   littoral drift
neap tide   barrier island
tidal bulge (tidal crest)   jetty
diurnal tide   dune
semidiurnal tide   groin

Now you are ready to begin the audiovisual portion of this module. Select the MEDIAPAK 3 component(s) and proceed. Following is a topical outline of the audiovisual sequence. You will find this outline helpful for reference and review. After completing MEDIAPAK 3, return to this book to perform the exercises.

OUTLINE

Waves

Factors in the generation of waves in the ocean

Wind waves

Capillary waves

Storm waves

Swell

Progressive waves

Parts of a wave

Orbital motions of water parcels in waves

Breakers

Breaker formation in the surf zone

Energy of breaking waves

> *Spilling breakers*
> *Plunging breakers*
> *Collapsing breakers*
> *Surging breakers*

Standing waves

Tides

Definition and measurement of tides

Generation of tides

> *Effect of sun and moon*
> *Displacement of the tidal bulge*

Effect of basin shape

> *Diurnal tides*
> *Semidiurnal tides*
> *Mixed tides*

Tidal currents

> *Reversing tidal currents*
> *Rotary tidal currents*

Shoreline Processes

Wave straightening of coastlines

> *Erosion*
> *Sedimentation*

Barrier-island formation

> *Littoral drift*
> *Migration of tidal inlets*

Beach processes

> *Effect of storms*
> *Role of dunes*
> *Beach-erosion control measures—groins and jetties*

Exercises

1. The three most important causes of waves are earthquakes, the gravitational attraction of the sun and moon, and _____.

2. As waves move out of a storm area, a sea is transformed into _____.

3. Wave energy is dissipated on the beach as turbulence, momentum of uprush, and _____.

4. *True or False:*
   A. Waves in shallow water become higher, steeper, and farther apart.
   B. Spring tides occur when the sun, moon, and earth are in line.
   C. The rotation of the earth displaces the tidal bulge in the direction of rotation.

5. Standing waves
   A. have a period determined by the size of the basin in which they are generated.
   B. in ocean basins have a period determined primarily by the strength of the tide-generating force.
   C. are found in the region around Antarctica.
   D. have their greatest horizontal movement at opposite ends of the basin in which they move.

6. Mixed tides
   A. have one high and one low tide per tidal day.
   B. are easy to predict because they tend to be the same height every day.
   C. have two unequal highs and lows per tidal day.
   D. are relatively rare over the earth as a whole.

7. Reversing tidal currents
   A. are found in open-ocean areas.
   B. ideally have their maximum speed shortly after high slack water and their minimum speed shortly after low slack water.
   C. are largely unaffected by winds and river discharge.
   D. are the strongest currents in many ocean regions.

8. Match the following:
   A. Smooth, rounded crests characterize
   B. The largest waves occur in ocean areas having a great

   1. two equal highs and lows per tidal day
   2. wave-straightened coastlines

C. Tides have some properties of

D. A basin that responds strongly to semidiurnal constituents of tide-generating forces is likely to have

E. In rotary tidal currents, there are no periods of

F. Inlets between barrier islands are kept open by

G. Movement of beach sand along the shore is called

H. Beaches are often rebuilt after a storm by

I. A continuing supply of sand contributes to

J. U.S. east and Gulf coasts commonly display

3. tidal currents
4. beach stability
5. normal wave action
6. littoral drift
7. shallow-water waves
8. distance across which winds can blow
9. swell
10. slack water

SUMMARY

Waves

Waves continuously cross the sea surface. They are generated whenever anything disturbs the water, such as a passing ship or a block of sediment sliding down the continental slope. The three most important generators of waves are the winds, the attraction of the sun and moon, and earthquakes.

Most of the ocean's waves are wind generated. Tiny ripples, or capillary waves, are set up by a light breeze forming more or less regular arcs. They may also form on top of preexisting waves. Ripples soon die out if the breeze drops, but under steady wind, they are gradually transformed into short, choppy waves.

As long as waves receive energy from the wind, they continue to grow larger. They form on top of other waves, seemingly moving in every direction. Under the influence of storm winds, a chaotic and complex mixture of different kinds of waves develops. This is known as a fully-developed sea.

The largest wind waves are formed by large, intense storms at sea. Three conditions cause unusually large, high-energy waves: high wind speed, wind blowing from one direction for a particularly long time, and the longest possible fetch, or distance across which a wind can blow without interruption. Ocean areas with the greatest fetch have the largest waves. For instance, the North and South Atlantic and Indian oceans have a maximum effective fetch of about 1,000 kilometers (600 miles). These basins rarely produce waves more than 15 meters (45 feet) high, whereas the Pacific Ocean, with a greater effective fetch, is known to produce waves at least 34 meters (100 feet) in height.

As storm waves move away from the winds that generated them, the chaotic sea surface becomes calmer. However, the energy imparted by winds has not been dissipated. The sharp-crested, irregular waves of the sea are transformed into long, smooth, regular swell that disperses from the generating area in a wide fan. Because little energy is expended as the rounded waves of a swell move over the ocean surface, they can travel virtually unchanged across hundreds or thousands of kilometers of open ocean.

Long-period swell (having an elapsed time of up to 16 seconds between the passage of successive wave crests) travels faster than short, choppy waves. Thus, long-period storm waves are the first to reach the shore, where energy originally imparted by winds is finally dissipated in the surf, forming breakers and moving beach sands.

Waves change shape when they enter shallow water. They become higher, steeper, and closer together. To understand why this happens, let us analyze the wave form, and how it travels through water.

A simple wave that moves in one direction only is called a progressive wave (Figure 3.1). Its height is the distance between the high point, or crest, and the low point, or trough. The still-water level would be midway between the two. Wave length is the distance between successive crests (or successive troughs). Wave period is the length of time elapsing between the arrival of two successive crests at a given point, for instance a pole in the water.

Figure 3.1 *Progressive wave.*

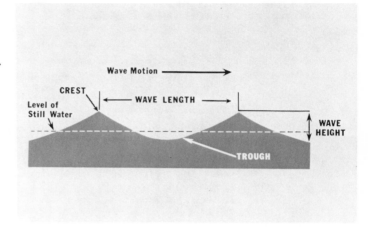

Although the wave form moves forward across the water surface, in a deep-water wave, there is virtually no net forward motion of the water itself. This can be seen by watching a chip of wood move on water as a wave goes by. The chip moves forward on the crest of the wave, but back to its original position in the trough. This happens because in deep water, particles move up and down in nearly circular orbits as a wave form goes by (Figure 3.2A). Orbit diameter is equivalent to wave height.

Below the surface, orbits become progressively smaller downward and nearly vanish at a depth of one half the length of the wave. If the water becomes increasingly shallow, so that the bottom is less than half a wave length below the surface, water-particle motion is lessened (Figure 3.2B). We might say that the wave "feels" the bottom. This causes the wave to assume the characteristics of a shallow-water wave. Water particles near the bottom can no longer orbit freely, causing the wave to travel more slowly, to become higher and steeper, and to decrease in length. Therefore, as swell nears the beach, and waves become higher, the breakers are closer together than were the crests of the swell that formed them. They also change direction so that the wave crests align themselves more nearly parallel to the depth contours of the bottom. This causes the wave crest to rotate and approach the beach directly, or at a slight angle. If wave approach is from an angle, sand is moved along the length of the beach in the direction the wave is moving. We will discuss this later, when we study beach processes.

When water depth is about one and a third times the height of a wave, the wave crest becomes unstable, and the wave starts to break (Figure

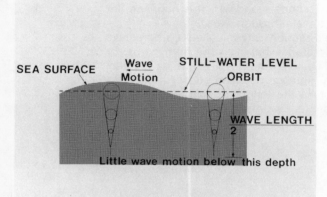

Figure 3.2 *Water-particle motion in (A) deep and (B) shallow water.*

A.

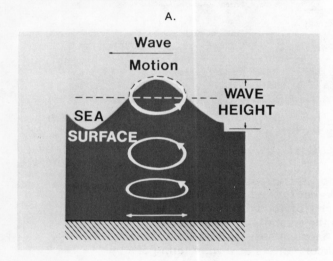

B.

3.3). Moving into increasingly shallow water, the wave re-forms and breaks again. Energy released by a breaking wave may cause a new set of smaller waves to form, and these also break when they reach shallower water. Thus, several sets of breakers may occur in a surf zone, depending on the amount of energy in the approaching swell, and on the shape of the bottom. Waves of different heights and speeds approach the beach from all directions. In addition, the force of breaking waves continually alters the configuration of the bottom. Thus, no two successive breakers form under exactly the same conditions, and no two are ever exactly alike.

In the final stages of a breaking wave, the water itself moves toward the beach. Part of the wave's energy is dissipated in water turbulence, and part is translated into the momentum that sends a thin layer of water rushing up the beach face. Some wave energy is changed into heat, which is absorbed by the large volume of water in the surf zone, and ultimately carried away from the beach. It has been estimated that if a wave 1.2 meters (3.5 feet) high, with a period of 10 seconds, were to strike the entire west coast of the United States, 50 million horsepower would be released.

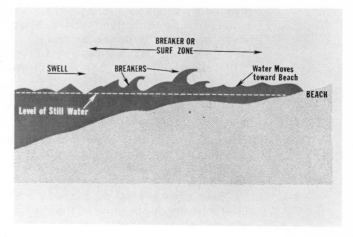

Figure 3.3 *Breaker formation.*

The shape of a breaker depends on its height and speed, which are in part determined by the slope of the beach and the amount of friction, or resistance, it offers to the approaching waves. Characteristics of breakers are summarized in Table 3.1.

TABLE 3.1   Types of Breakers and Beach Characteristics Associated with Each

Breaker type	Description	Slope and surface of beach	Water depth / Wave height
Spilling	Turbulent water and bubbles spill down front of wave; most common type	Gentle slope; irregular bottom	1.2
Plunging	Crest curls over large air pocket; smooth splashup usually follows	Moderately steep; smooth bottom	0.9
Collapsing	Breaking occurs over lower half of wave; minimal air pocket; usually no splashup; bubbles and foam present	Moderately steep	0.8
Surging	Wave slides up and down beach with little or no bubble production	Steep	Near 0

SOURCE: C.J. Calvin, "Breaker Type Classification on Three Laboratory Beaches," *Journal of Geophysical Research* 73 (1968): 3655.

In addition to the progressive waves we have been studying, there is another type of wave, known as a standing wave, in which the wave form does not move. A standing wave can be generated by tilting a round-bottomed dish of water, then setting it down. In the center of our small dish, or at various locations in an ocean basin, water has some horizontal motion, but no vertical movement. At opposite ends of our dish, or of an ocean basin, water moves up and down. There are none of the obvious orbital motions characteristic of a progressive wave. A standing wave's period is determined by the basin's dimensions.

Ocean Tides

Tides are long-period waves that cause sea level to rise and fall on most coastlines either every 24 hours 50 minutes, or every 12 hours 25 minutes. They are caused by the moon and sun, which attract the ocean's waters.

Each coastal area has a characteristic type of tide, which is slightly different every day because of the changing positions of the sun and moon with respect to the earth. The range, height, and period of tides are determined by these relationships but modified by the shape of the basin in which the tide moves.

Tide records for many coastal areas are kept on automatic digital computers. They show the changing level of the sea surface, which can be monitored by a float inside a wave-stilling pipe connected to the water. A tide curve is often drawn on a clock-driven paper, showing tide level plotted against time.

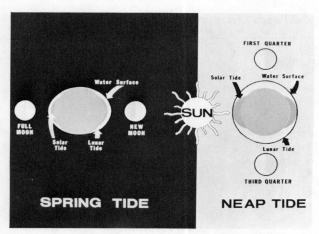

Figure 3.4 *Relationships of the sun, the moon, and the earth during (A) spring and (B) neap tides.*

A.   B.

We can visualize the effects of the sun and moon on tides by imagining the ocean as a watery envelope on the earth's surface. The envelope is mainly deformed by being drawn toward the moon, but the position of the bulge is also somewhat influenced by the sun's relative position. There are two tidal bulges at any time, on opposite sides of the earth. Each may also be thought of as a wave crest, having low tides, or troughs, occurring between the crests. The earth revolves beneath the moon once in every tidal day of 24 hours 50 minutes. When it is aligned with the sun, at times of the full and new moons, the highest tides, or spring tides, occur (Figure 3.4A). When the sun and moon counteract each other, as at the first and third quarters, there are lower or neap tides (Figure 3.4B).

If the ocean were infinitely deep, and if the earth was entirely covered by water and not rotating, high tides would always occur exactly at the point closest to the moon. In fact, the tide wave would follow the moon at a speed of about 1,600 kilometers per hour. But the tides have a wave length of 22,000 kilometers (13,660 miles), and the ocean is only about 4 kilometers (2.5 miles) deep. Thus, the tide behaves like a shallow-water wave, and its speed is reduced by friction from the ocean bottom. The earth's rotation introduces another factor, by displacing the tidal bulge in the direction of rotation. And in every part of the ocean, the tide wave responds to the shape of the basin in which it moves. So, the actual position of high tide in any part of an ocean basin is determined by an equilibrium among several factors, especially the position of the moon and the amount of friction experienced by the tide wave on the rotating earth.

We have already learned that the world's tides have some properties of a mammoth shallow-water wave. Around Antarctica, where no land interferes with its passage, the tide continually circles the South Pole as a forced wave. That is, it is generated and maintained by the moon's continuous attractive force rather than existing as a free wave after the generating force ceases. But in most parts of the world ocean, the tide has some characteristics of a standing wave in that it moves back and forth in each basin.

Every basin has its own natural period for standing waves. If that period is approximately 12 hours, or 25 hours, the standing wave component of the tide is likely to be well developed. If the basin's natural period is greatly different from 12 or 25 hours, the tide will behave less like a standing

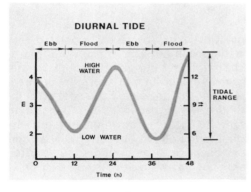

A.

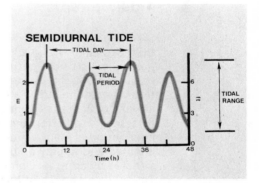

B.

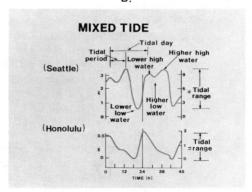

C.

3.5 *(A) Diurnal, (B) semidiurnal, and (C) mixed tides.*

wave. The Gulf of Mexico and parts of the Southeast Asian coast experience one high and one low tide per tidal day. This is the daily or diurnal tide characteristic of basins having natural periods of around 25 hours (Figure 3.5A).

The Atlantic Ocean tends to respond more readily to semidiurnal constituents of the tide-generating forces (Figure 3.5B). Therefore, two high and two low tides per tidal day are common. In this type of tide, successive highs and successive lows are roughly equal in height. The high tides usually occur at a regular time after the moon crosses a meridian of longitude, making these tides fairly easy to predict.

Mixed tides are most common over the earth as a whole. These have two complete tide cycles in each tidal day, but the two highs and the two lows are unequal in height (Figure 3.5C). These tides are difficult to predict because complex relationships exist between the tide-generating forces and the ocean's response to them. Basins such as the Pacific and Indian oceans and the Caribbean Sea respond to both diurnal and semidiurnal constituents of the tide-producing forces and so experience mixed tides.

Tides in bays and harbors are influenced by the size of the opening through which the tide wave enters from the open ocean. Coastal basins with small openings generally exhibit a smaller tidal range than are typical for less restricted areas.

Tidal Currents

Since tides are shallow-water waves, movements of water associated with their passage extend all the way to the ocean bottom. These water movements, called tidal currents, are the strongest currents in many coastal ocean regions. In coastal areas, their normal pattern is a reversal of direction after each slack water; that is, water floods toward the coast, or upstream, in an estuary until the high-tide level is reached. Then, the current reverses its direction and ebbs toward the ocean, or downstream, until the next period of slack water at low tide. Prediction of tidal currents is a complicated and largely empirical matter because these currents are so often affected by nontidal water movements such as wind effects and river flow. The time and speed of maximum flood or ebb current, which ideally occur midway between high and low water, may vary widely due to local conditions in a bay or harbor. Tidal-current tables predict these currents based on long series of observations.

Reversing tidal currents and their relationship to the stage of the tide in Chesapeake Bay are shown in Figure 3.6. These are relatively simple currents associated with a tide that resembles a progressive wave. We can follow their progress beginning with high slack water at the entrance to the bay (Figure 3.6A). From just north of this point, to low slack water about 120 kilometers (80 miles) up the bay, tide currents are ebbing. Still farther north, beyond the illustrated area, the currents are flooding toward another high-tide slack.

Two hours later, both areas of slack water have moved about 50 kilometers (30 miles) up the bay (Figure 3.6B). Note that at Norfolk, tidal currents reversed their direction after the flood tide moved upstream, as indicated by the arrows. Note also that the numbers show tidal currents to be moving at their maximum speeds of 2 or 3 kilometers per hour midway between slack-water areas. Near regions of high and low water, tidal cur-

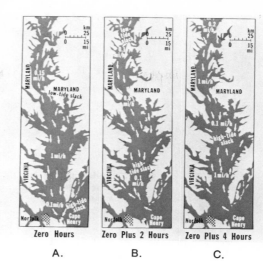

Figure 3.6 *Tidal currents in Chesapeake Bay.*

rents flow very weakly. After another two hours, the slack waters have moved still farther northward (Figure 3.6C).

In Long Island Sound (Figure 3.7), the tide behaves more like a standing wave. The entire sound experiences either ebb or flood at any given time. Water enters at The Race from the Atlantic Ocean, at the eastern end of the sound, and moves toward the western end, where Figure 3.7 shows slack water at high tide. Here, current speeds are given for spring tides. Note that tidal currents are strongest in the eastern end of the sound, where waters communicate directly with the Atlantic Ocean. They tend to be weaker in the interior of the basin.

Figure 3.7 *Tidal currents in Long Island Sound.*

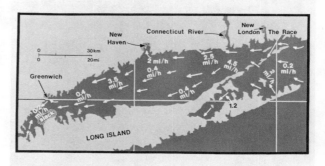

Tidal currents in open-ocean areas do not reverse direction as they do in coastal bays and harbors. Instead, they continuously change direction, moving clockwise, and there are no periods of slack water. At the end of each tidal cycle, the current is moving in the same direction as when the cycle began. For this reason, these are known as rotary currents (Figure 3.8).

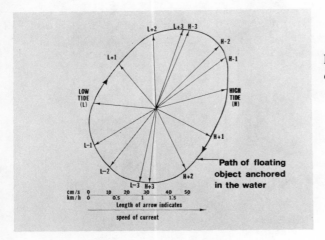

Figure 3.8 *Direction of rotary tidal currents off New England.*

Shoreline Processes

Continental shorelines are constantly worn down and reshaped by wave attack and scouring by tidal currents. These forces erode rough, irregular shoreline features, creating a wave-straightened coast. The coarse sands and gravels thus produced are deposited near the shore, forming beaches to

Figure 3.9 *Barrier islands at Cape Hatteras, North Carolina.*

which riverborne materials contribute in certain areas (for instance, on the U.S. West Coast). Fine sands and silts are transported onto the continental shelf. In many cases, eroded material is deposited across the mouths of bays and inlets, forming barrier islands, like those shown in Figure 3.9. These processes have only been acting on present shorelines for the past 3,000 years—since the ocean has been at its present level. In the future, bays and harbors will be filled in by riverborne sediment from continental interiors.

The long, narrow barrier islands that separate Pamlico Sound, North Carolina, from the adjacent North Atlantic Ocean were formed by waves and currents. Such islands are penetrated by narrow openings, or inlets, through which tidal currents carry ocean waters in and out of the lagoon. Inlets are moved by waves. New ones may be cut, or preexisting openings closed, during a single violent storm.

Normal wave action tends to close tidal inlets by moving sands along the coast. This movement of beach sand in caused by waves striking the shore at an angle rather than head-on and is known as littoral drift. Moving sands tend to be deposited in or very near the inlet, eventually closing it, or wave action may cause inlets to move "down-drift," that is, to move in the direction of wave-driven sands. Tidal currents have the opposite effect in that they act to scour loose material from tidal inlets, thus keeping them open.

The westward migration of Fire Island Inlet illustrates how a barrier island is extended by littoral drift (Figure 3.10). Fire Island is about 48 kilometers (30 miles) long and generally less than 1 kilometer wide. It is a typical barrier island of the type that extends along a large part of the U.S. east and Gulf coasts. Bays, marshes, or tidal lagoons separate such islands from the mainland, in this case, from Long Island, New York. (Long Island is an accumulation of rock and gravel deposited by a glacier during the ice ages.)

Barrier islands like Fire Island are made of sandy beach deposits that have accumulated over thousands of years. Along Fire Island's seaward edge, wave attack is not directly parallel to the beach but strikes at a slight angle. Sands are continually moved westward along the ocean

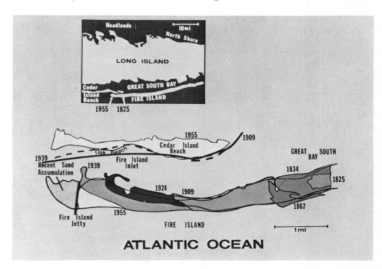

Figure 3.10 *Westward migration of Fire Island Inlet.*

beaches, at both Fire Island and Long Island itself. Because of this littoral drift, the western end of the barrier island migrated westward during the surveyed period from 1825 to 1939, at an average annual rate of about 70 meters (210 feet) (Figure 3.10).

In 1939, to prevent further westward migration, a low stone wall, or jetty, was built at the end of the island, extending southwestward. The 1955 survey showed that moving sands subsequently collected against the jetty and extended the beach seaward. Extension of the island and elongation of the inlet were temporarily halted. However, since 1948, sand has been spilling around the jetty, filling the inlet. Consequently, other problems have arisen. The inlet has shifted northward, and tidal currents are now scouring sand from Oak Beach (Figure 3.10).

Efforts have been made to rebuild Oak Beach by dumping more than three quarters of a million cubic meters (a million cubic yards) of sand on it, and by deepening the inlet close to the jetty. But because littoral drift continues to move vast amounts of sand westward, these measures have not yet solved the problem of how to stabilize sand movements in the area.

Marine processes do not effect such rapid changes in all types of coastlines. Granitic rocks, for instance, are very little changed on a short-term basis. Headlands of softer rock (see Figure 3.10) are gradually eroded, causing coarse beach deposits to accumulate nearby. Sands derived from coastal erosion are carried along the shore by currents and wave action. In quieter water, they settle out to form the sandy beaches we value so highly as recreational areas. This often occurs in the bays that separate headlands.

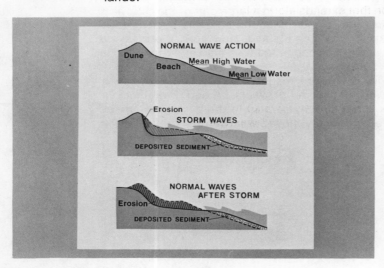

Figure 3.11 *Alteration of a beach profile by wave action.*

A beach is constantly shaped and reshaped by wave action. During calm weather, the beach-and-dune complex (shown in Figure 3.11) is generally quite stable, but during a stormy season, the shape of a shoreline may be markedly altered. Above the mean high-water mark, storm waves can cut into the beach and even erode the dunes beyond. The sand is washed offshore, where it often forms an underwater rise or bar.

Subsequent normal wave action, commonly during the summer season, tends to rebuild the beach. Erosion of a dune, however, takes longer to repair, because dunes are formed very slowly by sand blown back off the beach. They are kept stable by grasses that take root and spread over them, but this too is usually a slow process.

Protection of dunes is an important aspect of beach stability. These low hills of wind-blown sand act as barriers to storm waves that may otherwise flood and damage property farther inland. So in building beachfront houses, hotels, roads, and other structures, it is important to consider the effects that such construction will have on the beach as a whole and the processes that affect it. For example, leveling a dune to build a hotel may result in that hotel being destroyed in a particularly violent storm within a few years.

In planning for effective beach-erosion control, it is importnat to develop a program for an entire coastal area, because changes effected for a particular purpose locally may result in unexpected changes elsewhere down the coast.

A.   B.

Figure 3.12   *(A) Groins retard littoral drift caused by waves striking the coast obliquely. (B) Jetty near the mouth of the Columbia River.*

Low walls built seaward across beaches to retard littoral drift, called groins, are designed to slow down beach erosion by trapping sands that are carried along the shore by waves and longshore currents (Figure 3.12A). But material accumulating on the updrift side of a groin is no longer available to rebuild eroding beaches on the downdrift side. As a consequence, erosion below the structure is accelerated so that additional structures are required farther and farther down the coast.

Another way to deal with beach erosion is to add sand dredged from a channel, or brought from inland. We learned earlier that this procedure has been used on the barrier beaches that parallel Long Island's south shore, in New York State.

Tidal-current movements may also be altered by man-made barriers, for instance paired jetties (Figure 3.12B). A jetty is a low wall more massive than a groin which is built into the ocean at the entrance to a river or bay so that tidal currents are confined to a narrower channel. Tidal currents flow with greater force through the constricted opening, scouring out accumulated bottom deposits and acting to keep the inlet open. Jetties often trap sand moving in the littoral drift and form large beaches at the harbor entrance or river mouth.

Questions

1. What are three of the most important wave generators in the open ocean?

2. Describe the differences between capillary waves and the waves in a fully developed sea. How are the two kinds of waves related in time and space?

3. Explain how swell evolves from a fully developed sea, and how it moves through the ocean.

4. Draw a picture of a progressive wave, showing its parts.

5. What is meant by the term *wave period*?

6. Explain how a parcel of water moves as a wave passes in deep water. How is that motion different in a shallow-water wave? Draw a picture to illustrate the difference.

7. What causes a wave to break? Describe the changes in wave shape as a wave nears the beach. What finally happens to the energy of a wave?

8. How is a standing wave generated? What are its characteristics? Explain the motion of a standing wave in a basin of water.

9. What are the tide-generating forces? Explain how the alignment of the sun and moon produces the highest and the lowest tides. Name those two kinds of tides, and draw a picture to illustrate them.

10. How do the depth and shape of a basin govern the characteristics of tides? In a partially enclosed basin, what effect does the size of the inlet through which water enters have on the height of tides?

11. What is the effect of the earth's rotation on the position of the tidal bulge?

12. Explain the differences in height and timing between a diurnal tide, a semidiurnal tide, and a mixed tide.

13. Describe the stages of a reversing tidal current from low slack water through the entire tidal cycle to the next corresponding low slack water. Assume that you are observing the current from a fixed point on the shore.

14. What differences are there in a tidal current that acts like a progressive wave, as in Chesapeake Bay, and one that acts like a standing wave, as in Long Island Sound?

15. How is a rotary tidal current different from a reversing tidal current? Draw a picture to illustrate a rotary tidal current.

16. Explain the effect of wave action on an irregularly shaped coastline. Include an explanation of littoral drift.

17. Describe a barrier island and explain how the position of its inlets may be changed by wave action.

18. Describe the effect of winter storms on a beach. What effect does normal wave action have on a winter-eroded beach during less stormy seasons?

19. What is a dune, and what part does it play in stabilizing coastal areas?

20. Describe how a groin differs from a jetty, and explain the function of each.

Suggested Readings

Bascom, Willard. *Waves and Beaches.* Garden City, N.Y.: Doubleday & Company, 1964. *Elementary and engaging treatment.*

Russell, R.C.H., and MacMillan, D.H. *Waves and Tides.* London: Hutchinson, 1954. *Elementary discussion.*

MODULE 4

The Coastal Ocean

INTRODUCTION

Coastal oceans lie above continental shelves and contain only a small fraction of the earth's ocean waters. But it is that fraction which most directly affects human populations. Furthermore, the coastal ocean is strongly affected by winds, tides, and its own narrow boundaries, factors which play only a minor role in the open ocean. Therefore, it will be convenient to treat the coastal ocean in a separate section, detailing the conditions that make it a province with distinct oceanographic properties, and emphasizing its impact on humankind.

Coastal oceans and marginal seas cover 12.5 percent of the earth's surface, but because they are so shallow, they account for only about 4 percent of its volume. Throughout time, these highly productive waters have provided food for shore-dwellers and have served as a commercial highway. Even primitive boats can navigate many coastal regions. In the present era of extensive industrialization, nearshore waters have become even more vital to our welfare. A large percentage of the world's population lives within easy access of ocean margins. For example, in 1970, three quarters of the people in the United States lived in the 29 states that border either the ocean or one of the Great Lakes. It has been estimated that by the 1980s more than two thirds of that population will live in or near a large city. And seven of the largest cities in the United States lie near a sea-coast. Witness the growth of the largest urban regions—the Boston-New York-Philadelphia-Washington corridor on the east coast, and the San Diego-Los Angeles complex bordering the Pacific Ocean.

These large, concentrated populations place heavy demands on coastal-ocean resources. For example, coastal waters receive large amounts of waste material, both intentionally and by accident. Sewage sludge, construction debris, industrial acids, and other wastes are regularly disposed of at designated coastal-ocean dumping sites. In addition, erosion from suburban construction sites introduces large volumes of sediment into nearshore waters. Overflow from storm drains adds refuse from city streets, including wastes from domestic animals. Farming operations in river drainage basins that empty into coastal waters contribute residues from pesticides, fertilizers, and farm animal by-products. In the ocean itself, supertankers spill oil during offloading operations and occasional ship damage. Furthermore, dredging of navigation channels to permit passage of deep-draft vessels causes problems for bottom-dwelling marine animals, especially when the dredged material is dumped in another area, where it can smother organisms living on the bottom.

Sometimes the effects of man-made changes on marine environments are quite complex. For instance, coastal waters are highly productive of fish and shellfish, because a rich supply of phosphates and nitrates enters the ocean via river discharge. Coastal-ocean processes constantly recycle these *nutrients* to sunlit surface waters, where they support a relatively dense population of phytoplankton, the tiny floating plants that form the primary food source for all marine animal life.

Phosphates and nitrates are also discharged into coastal waters by sewage plants, cesspools, and fertilized farmlands. This causes an increase in phytoplankton populations, but it is not likely to cause a corresponding increase in fish or shellfish production. Rather, such overfertilization upsets the delicate balance between production and consumption of food in nearshore communities. Sometimes, there are not enough animals to eat the rapidly reproducing phytoplankton. The uneaten organisms sink to the bottom and die. They are then decomposed by bacteria. This process uses up the oxygen dissolved in deep waters, so the bottom becomes uninhabitable for most animal populations. Phosphates and nitrates released during the decomposition process are eventually recycled to the surface, where they again cause phytoplankton overproduction.

Furthermore, coastal circulation patterns act to retain the nutrients in nearshore waters. Thus, an overfertilized coastal area tends to remain out of balance for some time.

In general, much remains to be learned about how waste materials affect coastal waters, and how they are transported within that environment before being caught up in open-ocean circulation systems and thereby dispersed.

Damage to marine life through unsound waste-disposal practices becomes obvious to all of us when commercially harvested food sources are affected. But additional, often long-term degradation of marine environments has been noted by conservationists. Marshes and shoreline wetlands, known to harbor many species of young wild animals, fish, and birds, are often seriously altered by human use and abuse of the coastal ocean and its margins. Changes in salinity resulting from river modification or diversion of freshwater channels can affect local populations. Bridge construction often isolates sections of bays or lagoons, thus disturbing the ecological balance among plants and animals. Man-made temperature variations also have their effect. Waste heat from industrial plants or generators can kill fish.

On the other hand, young oysters have been grown in the heated waters from a power plant. It is possible that we can learn to improve the natural environment while at the same time putting it to use for our own needs.

Some coastal-ocean modification projects have been on a vast scale. For example, in the 1930s, the Zuider Zee, a large, shallow embayment bordering the Netherlands, was cut off from the North Sea by a dike. Within a few years, what had been a low-salinity estuarine area was transformed into a freshwater lake by pumping out the seawater and letting river water fill the area. Large sections were drained and converted to rich farmland (Figure 4.1). This project, which continues today as additional lands are prepared for agricultural use, is part of a centuries-old policy of land reclamation by the Dutch people. Although there have been substantial changes in local wildlife, the benefits to that densely populated nation have been very great. Careful management and use of natural resources often dictates compromise, and our view of what constitutes conservation must accommodate to the changing demands and stresses placed on the environment.

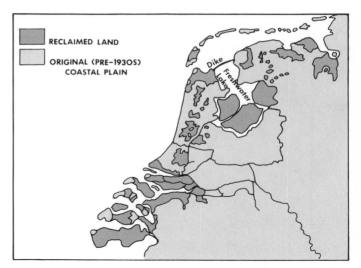

RECLAIMED LAND

ORIGINAL (PRE-1930S) COASTAL PLAIN

Figure 4.1 *Dutch lands reclaimed from the sea by dike construction and drainage.*

An understanding of oceanographic processes in coastal areas is important to effective planning for their management. In this module, we will study coastal-ocean circulation and learn how materials entering coastal waters from the continent move through nearshore waters in response to a combination of wind effects and density-controlled circulation patterns. *Estuaries* play an important part in coastal oceanography, so we will study estuarine processes in some detail. We will also study the effects of natural coastline evolution, during which estuaries and other bay and harbor areas are filled by entrapped sediment. In conclusion, we will briefly touch on a few presently available natural resources from the coastal ocean and examine the uses to which they are being.

OBJECTIVES

1. To be able to list some ways in which coastal-ocean processes differ from open-ocean processes

2. To be able to discuss the effects of winds on coastal circulation, and of storm winds in nearshore areas

3. To be able to describe basic estuarine circulation and list some of its results in estuaries and in the coastal ocean

4. To know two effects of tides on coastal circulation and mixing processes.

5. To understand why and how sedimentation occurs in estuaries, leading to marsh and delta formation

6. To know at least five resources of the coastal ocean, and to be able to approximate their relative value

KEY TERMS

coastal ocean
coastal current
storm surge
forced wave
estuary
salt-wedge estuary
estuarine circulation

tide rip
front
plume
delta
tidal marsh
nutrient minerals

Now you are ready to begin the audiovisual portion of this module. Select the MEDIAPAK 4 component(s) and proceed. Following is a topical outline of the audiovisual sequence. You will find this outline helpful for reference and review. After completing MEDIAPAK 4, return to this book to perform the exercises.

OUTLINE

Introduction to the Coastal Ocean

Rapidity of change; extremes of temperature and salinity

Vertical structure

Summer

Winter

Dimensions

Wind-influenced Coastal Circulation

Geostrophic currents

Upwelling

Piling up of low-density coastal water against coasts

Storm surges

Development within a coastal basin

Effects on shorelines

Estuarine Processes

Introduction to estuaries

Definition

Location in the United States

Formation

Estuarine circulation

Salt-wedge estuary

Moderately stratified estuary

Tidal effects

> *Evalution*
> *Tide rips*
> *Control by tides of freshwater entry to continental shelf*
> *Mixing of low-salinity waters on shelf*

Estuarinelike circulation in the ocean

Sedimentation

> *Retention of sediment by estuarine circulation*
> *Marsh formation*
> *Delta formation*

Coastal-ocean Resources and Their Value

Population pressures on coastal-ocean and shoreline resources

> *Shipping*
> *Sand and gravel*
> *Construction of offshore facilities*
> *Waste disposal*

Food and recreation

> *Fishing and mariculture*
> *Swimming, boating, and sportfishing*

Coastal engineering

Mineral resources

> *Metals*
> *Petroleum and natural gas*

Exercises

1. *True or false:*
   A. Large temperature changes are less common in coastal waters than in the open ocean.
   B. Geostrophic currents are caused solely by wind effects.
   C. A storm surge can set up a standing wave.
   D. Most people come in contact with the coastal ocean through recreation.

2. The coastal ocean's biggest industry in terms of dollar value is _____.

3. Two kinds of wastes commonly disposed of at sea are _____ and _____.

4. By 1985, 20 percent of U.S. _____ production may come from the coastal ocean.

5. Sand and gravel are commonly used for _____ in adapting shoreline areas for residential, commercial, or industrial use.

6. In a moderately stratified estuary
   A. sediment tends to be carried out of the estuary to the open ocean.
   B. a relatively large amount of subsurface water is entrained in the seaward-moving upper layer.
   C. a sediment-laden plume of high-salinity water at the surface often extends hundreds of kilometers along the coast.
   D. relatively little subsurface water is entrained in the seaward-moving upper layer.

7. In a salt-wedge estuary
   A. there is relatively little mixing between the layers.
   B. the pycnocline is not well developed.
   C. mixing of river water with seawater takes place entirely in the estuary.
   D. a large amount of seawater enters the estuary at the surface to replace the amount carried seaward along the bottom.

8. Where a large river flows through a small estuary
   A. fresh water has a long residence time in the estuary.
   B. mixing of river water with seawater takes place primarily on the continental shelf.

C.   estuarine circulation causes sediment to be deposited in a shallow area near the head of the estuary.

D.   moderately stratified estuarine circulation is likely to occur.

9.   Match the following:

A.   Geostrophic coastal currents result from a(n)

B.   Phytoplankton grow in the

C.   High tides in coastal areas can be associated with a(n)

D.   Partial mixing of fresh and salt water within an estuary is characteristic of a(n)

E.   An estuarinelike coastal circulation may result in

F.   Delta formation is associated with a(n)

G.   After an estuary is filled with sediment, it forms a(n)

H.   Additional food from the coastal ocean may be produced through

I.   A salt-wedge estuary has a(n)

J.   An estuarine delta is formed by

1.   weak tidal-current system near the river mouth

2.   sedimentation near the head of the estuary

3.   storm surge

4.   elevated sea surface due to low-salinity water piled up along a coast

5.   moderately stratified estuary

6.   sunlit surface layer

7.   tidal marsh

8.   retention of solid wastes near the coast

9.   mariculture

10. well-developed pycnocline

SUMMARY

Coastal oceans lie over the world's continental shelves. Their average water depth is about 70 meters (200 feet), with a maximum of approximately 200 meters (600 feet). This is in contrast with the open ocean, where water depths range from 4,000 meters (2.5 miles) to 6,000 meters (4 miles). Where the continental shelf is narrow, as on the Pacific coast of the United States, characteristic coastal-water properties may extend somewhat seaward of the continental shelf.

Coastal oceans are bounded by the adjacent continent and by the shallow ocean bottom. These have distinct effects on oceanic processes. Other factors that exert more influence in coastal oceans than in the deep ocean are tides, winds, and river discharge. Because coastal areas contain less water, large changes in water properties occur relatively rapidly. In a matter of days, or even hours, the coastal ocean can experience marked heating by the sun, cooling by radiation, dilution by rainfall and river discharge, or increased salinity due to excess evaporation. Extremes of temperature and salinity also characterize various parts of the coastal ocean. Such extremes may be noted on the surface-temperature and salinity profiles for the world oceans given in Figures 2.6 and 2.7.

At low and mid-latitudes, where river discharge and rainfall are moderate to high, and net warming occurs during all or part of the year, a warm, low-salinity surface-water layer is typical. A well-developed pycnocline prevents this layer from mixing with deeper waters. During mid-latitude winters, however, surface waters are cooled, and storm waves mix the water column from top to bottom. The pycnocline is destroyed, and deep waters are brought to the surface. This is an important factor in the biological productivity of coastal waters, because the deep layer contains nutrients that have been removed from the surface waters during the previous season. When nutrients are mixed to the surface, they become available to the new crop of phytoplankton that will bloom in the stable, sunlit surface layer during the following spring and summer.

Coastal-ocean processes are generally confined in space as well as in time. Marginal ocean basins, bays, and gulfs tend to have limited communication with the open ocean, so their properties are not dissipated by mixing with large amounts of seawater. Temperatures and salinities may vary within a few tens of kilometers, a rare occurrence in the open ocean. General characteristics of coastal oceans and the open ocean are shown in Table 4.1.

Let us begin our study of specific coastal processes with an examination of wind effects, which are commonly experienced in coastal regions within limited time and space.

TABLE 4.1 Comparison of Coastal and Open Oceans

Feature or process	Coastal ocean	Open ocean
Ocean bottom		
Depth	200 meters (600 feet) or less	Typically 4,000–6,000 meters (2.5–3.6 miles)
Influence	Significant influence	Insignificant except on near-bottom currents
Temperature-salinity	Large seasonal changes	Small seasonal changes
Currents	Generated by winds, tides, major ocean currents	Wind-driven and geostrophic currents
Surface currents	Parallels coastlines	Nearly east-west
Subsurface currents	Often oblique to coast, as in estuarine circulation	Nearly north-south in deep-ocean basins
Replacement of waters	Typically 1 year or less	Surface layers, 10–25 years; Subsurface waters, 250–1,000 years
Biological productivity	Relatively large	Relatively small, except along equator
Suspended particles	Relatively abundant, especially near river mouths	Relatively scarce

Coastal Currents

Coastal circulation is strongly affected by winds. In a previous section we discussed upwelling, during which a wind blowing toward the equator and parallel to the west coast of a Northern Hemisphere continent drives surface waters seaward. This permits deep, cold, nutrient-rich water to rise to the surface, causing climatic changes and increased biological productivity in the region.

Another important wind effect occurs when low-density surface waters are blown toward the coast instead of away from it. Figure 4.2 shows coastal currents flowing parallel to continental shores. These develop when low-density water, diluted by river discharge, piles up along the coast and depresses the pycnocline. The resulting low-density "hill" sets up geostrophic currents that flow parallel to the coast. These currents are strongest when river runoff is large and winds strong and steady. They diminish or disappear during seasons of light or variable winds and reduced freshwater discharge. Beyond the continental shelf, boundary currents flow steadily throughout the year, forming a barrier that separates geostrophic coastal circulation from the open ocean (Figure 4.2).

Figure 4.3 shows winter and summer circulation patterns off the Washington-Oregon coast, in the Northeast Pacific coastal ocean. During the winter, numerous coastal streams discharge into the ocean, and rainfall is abundant. A band of low-salinity water forms and is held near the coast by winds from the southwest. (Recall that this situation is opposite the upwelling conditions of summer, when winds from the north drive surface waters offshore.) The low-salinity water near the shore forms a

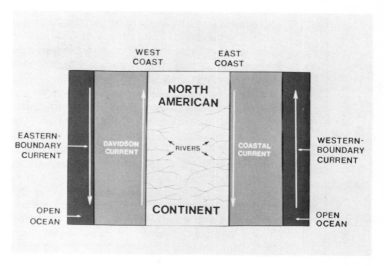

Figure 4.2 *Geostrophic coastal currents.*

seaward-sloping "hill" along the coast. A balance between gravitational attraction, directing surface waters downhill, and clockwise deflection due to the Coriolis effect gives rise to the northward-flowing Davidson Current. This current develops during the winter and continues to flow northward throughout the early spring. During summer and autumn, rainfall and river discharge are low along the Pacific Northwest coast. The Davidson Current diminishes, then disappears.

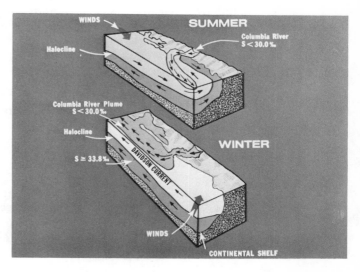

Figure 4.3 *Northeast Pacific coastal-ocean circulation.*

The California Current flows toward the equator throughout the year. It is the eastern-boundary current of the North Pacific open-ocean current gyre, and is located offshore, beyond the coastal circulation. During the winter, it moves in a direction opposite the Davidson Current, as we see in the diagram of generalized Northern Hemisphere circulation near North

America (Figure 4.2). During the summer, however, the typically weak, diffuse flow of this eastern-boundary current impinges on continental-shelf waters. The low-salinity plume of Columbia River water, which provides a relatively large freshwater discharge even in summer, is carried southward by the California Current (Figure 4.3). Below the pycnocline, a counterflow carries subsurface waters in the opposite direction. Note also in Figure 4.3 that summer winds blowing from the north drive the Columbia River plume somewhat seaward, creating upwelling conditions adjacent to the coast.

Geostrophic coastal circulation paralleling the coast acts as a barrier to water movements between continental margins and the deep ocean beyond. Water discharged by rivers or derived from rainfall along the coast must cross the coastal currents to mix with open-ocean waters. This transfer of fresh water is often quite slow. For example, off the Mid-Atlantic coast of the United States between Cape Cod, Massachusetts, and Cape Hatteras, North Carolina, coastal waters have a lowered salinity that represents two and a half years worth of accumulated discharge from regional rivers. We say that water in this coastal area, which is known as the New York Bight, has a residence time of two and a half years. By comparison, residence time for fresh water in Delaware Bay is only three or four months, and the Columbia River estuary, because of its small size and large river flow, has a residence time of about one day.

Long residence times have serious implications for highly industrialized areas, like the New York Bight. Wastes discharged from New York Harbor do not mix readily into the deep ocean but are retained in the shallow waters over the continental shelf. Materials released offshore of one harbor may wash ashore on beaches down-current and eventually damage their recreational potential.

In addition to setting up seasonally changing circulation patterns, wind effects can also be felt intermittently in coastal waters as a result of storms. During a storm surge, strong winds can cause water to pile up against a coast, creating an extremely high sea level in certain areas. Sometimes, an unusually high wave created by storm winds can travel as a forced wave with a gale wind for a long distance across the open ocean. When such a high wave enters a partially enclosed area, it often sets up a standing wave, which may oscillate in the basin after the storm has passed, as was the case in the 1953 North Sea storm surge, whose tidal effects are shown in Figure 4.4.

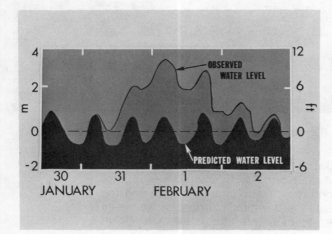

Figure 4.4 *Tidal curve on the Netherlands coast, January 30–February 2, 1953.*

This disastrous storm surge was set up by a northwest gale blowing at a rate of 25 meters (75 feet) per second across the North Sea, which has a fetch of about 900 kilometers (500 miles). It caused sea level to reach a height of 3.3 meters (about 10 feet) on the Dutch coast, which was more than 2 meters (6 feet) higher than the predicted high-tide level. Waves broke through the dikes and dunes that protect the low-lying Dutch coast, causing extensive destruction of property and loss of life inland.

Estuarine Circulation

Much of the water that drains from continents, which carries suspended and dissolved materials, enters the coastal ocean through estuaries. An estuary is a semienclosed coastal region where seawater mixes with the fresh water from land. On coastal plains, most of these embayments occupy ancient river valleys. Lagoons, separated from the ocean by barrier beaches, are also estuarine systems. Estuaries, lagoons, and deltas make up 80–90 percent of the North American Atlantic and Gulf coasts. The West Coast has fewer such systems, because its mountainous topography does not provide many passages to the ocean. Because estuarine processes control the rate at which fresh water and suspended materials enter the adjacent ocean, we devote considerable space to them in this section.

Modern estuaries were formed during the most recent rise in sea level. Fifteen thousand years ago, sea level was about 125 meters (400 feet) lower than at present, and the continental shelf as we know it was a coastal plain. Narrow river channels cut across the plain, draining water to the ocean, whose shores lay near the edge of present continental shelves. After the last ice age, sea level rose rapidly. For 6,000 years, it advanced across

Figure 4.5 *Chesapeake Bay estuarine system.*

the continental shelf, inundating river valleys. Estuaries were progressively displaced inland, so rapidly that sediment from the continents was insufficient to fill them. Over the past 9,000 years, sea level has risen only about 20 meters (60 feet); many small estuaries have been completely filled during this period of slowly rising water. Chesapeake Bay, shown in Figure 4.5, is one of the largest estuarine systems in the United States. Many rivers drain into it, including the Susquehanna, the Rappahannock, the York, the James, and the Potomac. Delaware Bay, another large estuary, lies to the north.

An estuary acts as a two-way street for moving water. River water flows in at its head (also at its sides if many rivers discharge into it). This less dense fresh water does not immediately mix with the seawater in the estuary's deeper layers. Rather, it spreads out and moves seaward as a shallow surface layer. If the estuary is relatively deep, with a large river flowing into it, the fresh water is often separated from the saltier, denser water below by a well-developed pycnocline. In this type of system, known as a salt-wedge estuary (Figure 4.6), there is relatively little mixing between the layers. However, because of frictional drag between the seaward-moving fresh water and the saltier water below, some seawater is entrained (dragged along) by the upper layer and transported seaward. The incorporation of deeper water into the surface layer increases the volume of water moving out of the estuary and also makes it saltier.

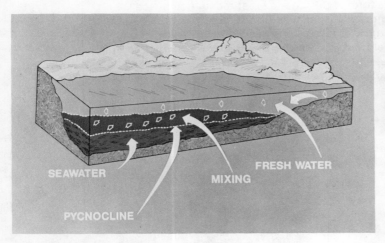
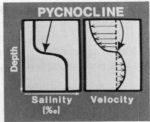

Figure 4.6   *A typical salt-wedge estuary.*

In a typical salt-wedge system the seawater layer becomes shallower toward the head of the estuary, forming the wedge shape that gives it its name. The lower Mississippi River and the Columbia River during flood stages exhibit this type of circulation, with current speed being greatest at the surface, least at the pycnocline. This is characteristic of fairly deep estuaries where tidal range is low and river discharge heavy.

Compared with the typical salt-wedge estuary, a much larger volume of water is entrained in the seaward-flowing upper layer of a moderately stratified circulation system. As a result, the volume of water flowing out into the open ocean is much greater than the amount brought in by rivers. Figure 4.7 shows that if a volume of water (R) enters at the head of an estuary, as much as 10R may be entrained from deeper layers, to be carried out to the coastal ocean. To replace this large volume of water lost at the surface, a comparable amount moves in near the bottom. Thus, the net effect of estuarine circulation is to draw seawater toward the coast.

Figure 4.7 *Relative amount of in- and outflow in a moderately stratified estuary.*

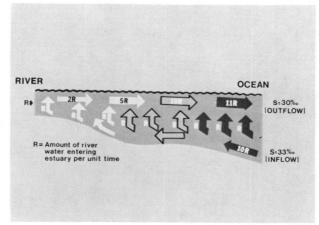

Tides in Estuaries

Tides dominate the circulation in most estuaries. Any current visible to an observer on the bank is most likely tidal in origin, and not an obvious result of estuarine circulation. To observe the nontidal component, we can measure the flow with a current meter and average the ebb and flood currents over an entire tidal cycle. The two cancel each other out, since the same amount of water enters the estuary on the flood tides as leaves it on ebb tides. Any inequality is due to estuarine circulation. We can also observe this by timing the duration of inflowing and outflowing currents. In the open ocean, these would be nearly equal, but in an estuarine system, the ebb currents are stronger and of longer duration at the surface than are flood currents, because most of the water volume gained from rivers leaves the estuary in surface layers. Flood currents are strongest in deeper layers.

Tide rips may form in an estuary where tidal effects are pronounced, as shown in Figure 4.8A. Here, we see the tide beginning to rise, with seawater moving landward below the downstream flow of the river. At first, just after low slack water, the flood current is much weaker than the river flow. But the flood current gains strength until a point midway between low slack water and high slack water, when tidal currents are typically strongest. At the same time, the river flow becomes progressively weaker, because the flooding tidal current is carrying it backward, just as a person running down an 'up' escalator makes less and less downward progress if the escalator accelerates as he runs. Somewhere before the midpoint is reached, the tidal current gains sufficient speed to be running upstream exactly as fast as the river water is flowing downstream. At that time, relative downstream motion in the surface layer ceases, and a convergence, or tide rip, forms where the two layers have met (Figure 4.8A). For a while, the river seems to reverse itself and run upstream, carried by the force of the flooding tidal current below. But as slack high water approaches, the force of the flood current diminishes, so the river again acquires down-stream momentum (Figure 4.8B). The rip then moves toward the sea and is carried out of the estuary on the ebbing tide.

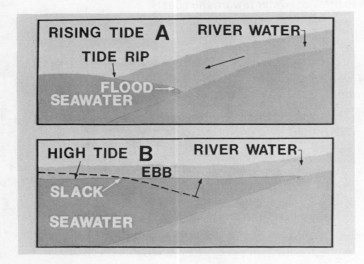

Figure 4.8   *Formation of a tide rip.*

Thus, the rise and fall of tides tends to control the timing of river discharge into the coastal ocean. Water from the estuary moves into the ocean as a series of pulses, one on each ebb tide. Seen from the air, these appear as a sequence of cloudlike water parcels, overlapping and becoming more faintly outlined with increasing distance from the river mouth. Most of their cloudlike appearance is due to fine sediment being carried out of the estuary. Consequently, discharge from small, sediment-filled estuaries is usually more visible than discharge from large, deep estuaries, where sediment is normally trapped by the estuarine circulation. Chesapeake Bay discharge, for example, does not usually contain much sediment.

Coastal Mixing Processes and Sedimentation

Where a large river flows through a small estuary, as is the case for the Columbia River, mixing of river water with coastal water occurs on the continental shelf rather than in the estuary. Such mixing takes place at the boundaries of the discharged, low-salinity water masses. Boundaries often remain sharply marked at zones of intensive mixing. The noise made by choppy, turbulent waters at such a boundary, or front, can sometimes be heard by shipboard observers as the front moves through calm coastal waters.

As overlapping low-salinity water parcels move farther away from the river mouth, the boundaries between them are gradually obscured, and a low-salinity plume of discharged waters is formed offshore. Such a plume is often identifiable for tens or hundreds of kilometers as it moves along the coast, held there by coastal circulation (see Figure 4.2). Eventually, waves and tidal currents mix this water with adjacent and underlying coastal-ocean waters, causing it to lose its distinguishing characteristics.

Estuarinelike circulation is not restricted to embayments or river mouths. Upward movement of bottom waters into a less-saline surface layer is rather common in coastal areas, where it tends to cause movement of bottom materials shoreward. It also occurs below the surface layer in large parts of the open ocean, for example the North Pacific and northern Indian oceans, and the Arctic Sea.

Estuarine circulation causes sediment to be retained in estuaries (Figure 4.9). Swiftly flowing rivers carry sand and silt-sized particles from the continental drainage basin. Gravel-sized material tends to roll down the river bed, propelled by the force of the water. Upon reaching the head of an estuary, where the channel widens, river flow loses much of its momentum. Coarse-grained sediment settles out at this point and is deposited in the estuarine delta that forms a shallow region at the head of the estuary (Figure 4.9). Finer-grained material remains suspended in the surface layer and is carried farther out into the estuary. There it gradually settles into deeper waters and is thus caught in the subsurface landward flow. Therefore, rather than being carried out with the ebbing tide, most of the sediment load that enters a deep estuary is deposited within its basin. In a shallow estuary, estuarine circulation tends to be absent, or very little developed, so sediment is carried out into the coastal ocean.

Where rivers carry heavy sediment loads, an estuary fills rather rapidly. The estuarine delta becomes enlarged and extends seaward. Sedimentation occurs in bays and marginal areas where tidal currents are less strong. As a consequence, marshy regions develop along the estuary's banks, and the sea grasses that take root in the soft, uncompacted bottom deposits act to further trap sediment.

Figure 4.9 *Sedimentation due to estuarine circulation.*

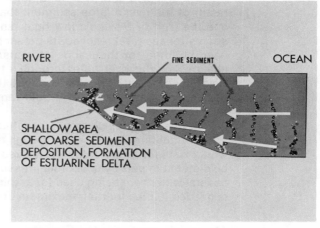

If an estuary is not kept open by regular dredging, river flow will eventually be confined to a few winding channels in a tidal marsh (Figure 4.10). The surface of such a marsh is flooded by high tides but protected from direct wave attack because of its sheltered position. The river continues to supply this region with nutrient-rich sediment, so salt-tolerant vegetation grows abundantly. Young fish, water birds, and other coastal wildlife find food and refuge there. As a result, tidal marshes are highly valued for the variety and profusion of their plant and animal life.

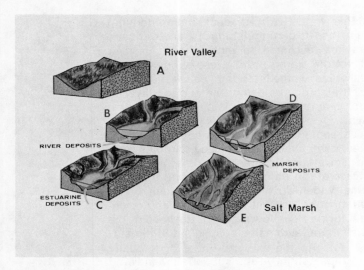

Figure 4.10   *Sedimentation of an estuary.*

As a consequence of estuarinelike coastal-ocean circulation, most riverborne sediment discharged to nearshore waters is held near the coast by landward flow in the subsurface layer. Geostrophic currents running parallel to the shore have the same effect. Where tidal and longshore currents are weak, or where coastlines are protected from strong wave attack, sediment tends to settle near river mouths rather than being dispersed along the coast. Deltas sometimes form as a result, especially in rivers whose sediment load is unusually large, like the Nile River in Egypt.

The Mississippi River Delta, largest in the United States, formed as a result of the river's large sediment load (about 300 million tons per year) and the Gulf of Mexico's low tidal range. Its estuary was probably filled soon after sea level reached its present position, and its present delta has been building for thousands of years. A complex, branching system of channels carries river water through it to the Gulf of Mexico. The Columbia River, by contrast, discharges about 10 million tons of sediment into the ocean every year, but a large tidal range and strong waves prevent delta formation.

In this module, we have presented a brief survey of coastal and estuarine circulation processes. Although much has yet to be learned, a framework exists for future study. New directions in coastal-ocean research will be dictated in part by economic pressures. We will need to know what our coastal waters are being used for at this time, and what further information is needed to plan for their maximum utility in the future.

Uses of the Coastal Ocean

The density of populations concentrated along ocean margins, and the need for a constant supply of goods, result in the coastal ocean's biggest industry—transportation. Every year, many billions of dollars are spent on shipping in continental-shelf regions of the United States. For this use, water quality and coastal circulation processes are not important, but dredging of navigation channels can create problems.

To reduce the cost of channel construction and maintenance, a knowledge of local sedimentation processes is helpful. For instance, channels that are dredged in areas of rapid sedimentation will soon fill up again, and dredge spoils that are dumped near a newly deepened channel are likely to be carried back into the channel by bottom-current movements.

Other valuable resources related to shoreline population pressures are sand and gravel. These materials are used for making cement and as landfill in marshes and the edges of bays, where undeveloped land is being prepared for residential, commercial, or industrial use. Although sand and gravel have usually been taken from continents, offshore production is becoming more important. Airport and harbor facilities are often built on fill land, and it is likely that the number of offshore sites will increase. Offshore sites will be especially suitable for deep-water port facilities, for industries that dissipate large amounts of waste heat, and for installations whose operation might threaten the safety of nearby populations.

The coastal ocean is used extensively for waste disposal, especially near large cities. Sewage, industrial chemicals, construction debris, and dredge spoils are the most abundant materials commonly handled in this way. It is difficult to assign a dollar amount to the coastal ocean's value as a dumping ground, but that value is at least as great as the cost of getting rid of the wastes in some other way. Further research in connection with the disposal industry will include studies of coastal circulation patterns, both near the surface and at depth. For instance, estuarine circulation in the New York Bight (Figure 4.11) tends to move solid waste materials landward, winds hold surface waters along the coast, and geostrophic currents prevent wastes dissolved in near-shore waters from mixing readily into the deep ocean.

Figure 4.11 *Dumping sites in the New York Bight.*

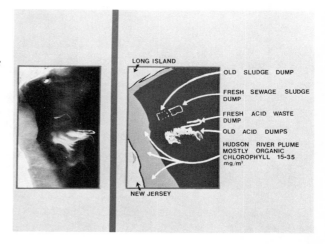

The coastal ocean contains most of the world's largest fisheries because of the high nutrient content of its waters and the tendency of bottom layers to be mixed upward and shoreward. The food industry in U.S. continental-shelf waters is worth several hundred million dollars per annum. To protect, and perhaps augment, this important resource, an understanding of water movements and water quality is useful. It is also important to understand the part marshes and other shoreline features play in the life cycle of young food organisms. Finally, the coastal ocean's potential for mariculture, or domestication of marine fish and shellfish, offers interesting possibilities for the future.

Billions of dollars are spent every year in recreational use of the ocean, including swimming, sailing, and sportfishing. For this industry, water quality and conservation of shoreline resources are vital. Lying on an oil-covered beach or fishing in polluted waters has little appeal for vacationers. Probably more people come in contact with the ocean through leisure-time activities than in any other way.

Coastal engineering is another major industry, valued at a few billion dollars annually in the United States. Construction of shoreline facilities such as harbors, and structures for protection of shorelines such as sea-walls and groins, fall under this heading, as do offshore structures like oil-drilling rigs. In order for these facilities to operate effectively, they must resist erosion and wave attack. Coastal engineering research continues in these fields, and much remains to be learned.

Coastal mining and petroleum production together are valued at a few billion dollars. Recoverable minerals on the continental shelf include tin, sulfur, gold, diamonds, and certain heavy minerals used in industry. For North America, the continental shelf is the largest unexplored source of natural gas and petroleum. It has been estimated that by 1985, 20 percent of U.S. petroleum production, and 30 percent of the gas, will come from the shelf. Seawater itself is most valuable as water for drinking in arid climates. Production of salt by evaporation is a minor, though viable industry in the United States.

Questions

1. Name three factors that play larger roles in coastal-ocean processes than in open-ocean processes.

2. Do coastal-ocean properties tend to change more rapidly or less rapidly than open-ocean properties? Explain your answer.

3. Is a pycnocline a typical winter feature in mid-latitude coastal oceans? Why or why not?

4. How are geostrophic currents set up in the coastal ocean?

5. The northeast Pacific coastal ocean receives more rain and river run-off in winter than in summer. Explain what effect this climatic change has on coastal circulation.

6. Explain what is meant by *residence time.* Name two factors that can affect residence time in a coastal-ocean area, and explain the effect of each.

7. Describe some effects that a storm surge can have on a coastal area.

8. Define *estuary.*

9. Describe salt-wedge stratification in an estuary.

10. How does a moderately stratified estuary differ from a salt-wedge estuary?

11. Describe one way of measuring the effects of estuarine circulation on water flow in an estuary.

12. What is a tide rip? Describe the conditions under which it can move landward in an estuary.

13. In what way can tides affect the rate at which river water enters the coastal ocean?

14. What two characteristics of an estuarine system might cause it to exhibit little mixing of seawater with fresh water in an estuary?

15. What coastal-ocean processes cause mixing of fresh water and sea-water on the continental shelf?

16. Explain what is meant by estuarinelike coastal-ocean circulation. How does it affect movement of materials on the bottom?

17. How does estuarine circulation cause sedimentation in estuaries?

18. Describe a tidal marsh.

19. Describe a delta. How is one formed?

20. Name at least six industries that are dependent on the coastal ocean and adjacent shorelines.

21. What is the coastal ocean's biggest industry in monetary terms? What changes in the coastal-ocean environment may result from the requirements of this industry?

22. What problems associated with waste disposal in coastal waters can be caused by the following?
    A. Direct wind effects.
    B. Geostrophic currents.
    C. Estuarinelike circulation

Suggested Readings

Committee on Resources and Man. *Resources and Man: A Study and Recommendations.* San Francisco: W.H. Freeman & Company, 1969. *Includes analysis of the resource potential of the ocean.*

Shepard, F.P. *The Earth beneath the Sea.* Rev. ed. Baltimore: Johns Hopkins Press, 1967. *Discussion of coastal processes and features.*

Skinner, B.J., and Turekian, K.K. *Man and the Ocean.* Englewood Cliffs, N.J.: Prentice-Hall, 1973. *Elementary treatment of oceanography and man's impact.*

Answers to Exercises

Module 1

1. Northern; **2.** Antarctica; **3.** Atlantic; **4.** A. False, B. False, C. True; **5.** D; **6.** B; **7.** A; **8.** A. 5, B. 4, C. 1, D. 10, E. 9, F. 7, G. 3, H. 2, I. 6, J. 8

Module 2

1. D; **2.** C; **3.** B; **4.** B; **5.** Salt; **6.** Over, above; **7.** Gyre, current gyre; **8.** A. False, B. False, C. True, D, True; **9.** A. 10, B. 13, C. 14, D. 3, E. 11, F. 6, G. 7, H. 5, I. 9, J. 2, K. 1, L. 8, M. 12, N. 4

Module 3

1. Winds; **2.** Swell; **3.** Heat; **4.** A. False, B. True, C. True; **5.** A; **6.** C; **7.** D; **8.** A. 9, B. 8, C. 7, D. 1, E. 10, F. 3, G. 6, H. 5, I. 4, J. 2

Module 4

1. A. False, B. False, C. True, D. True; **2.** Transportation; **3.** Sewage sludge, construction debris, dredge spoils, industrial chemicals; **4.** Oil, natural gas, petroleum products; **5.** Landfill; **6.** B; **7.** A; **8.** B; **9.** A. 4, B. 6, C. 3, D. 5, E. 8, F. 1, G. 7, H. 9, I. 10, J. 2

glossary

Antarctic Bottom Water   The densest water mass in the ocean; is formed near Antarctica, where it sinks to the bottom and spreads northward in all three ocean basins.

Antarctic Convergence   A region of sinking surface waters occurring between Southern Hemisphere current gyres and the West Wind Drift around Antarctica.

Antarctic Intermediate Water   A dense water mass that sinks at the Antarctic Convergence and forms a layer above the Antarctic Bottom Water.

antinode   The part of a standing wave where vertical water motion is greatest.

atoll   A ring-shaped reef enclosing a shallow lagoon and surrounded by open sea.

barrier island   A bar parallel to the shore, usually separating a shallow lagoon from the coastal ocean.

barrier reef   An reef separated from a landmass by a shallow lagoon.

breaker   A wave breaking on the shore or over a reef, etc.

capillary wave   A small, regular wave or ripple, usually caused by a light wind.

coastal current   A current flowing parallel to the shoreline.

coastal ocean   The relatively shallow waters overlying continental shelves, generally 200 meters (600 feet) or less in depth.

continental crust   The thickened, granitic part of the earth's crust that forms the continental blocks.

continental margin   The submerged edge of a continental block, consisting of continental shelf, slope, and rise.

continental rise   A generally smooth, gently sloping region that rises from the deep-ocean floor to the base of the continental slope.

continental shelf   The relatively shallow sea floor adjacent to a continent, overlain by the coastal ocean.

continental slope   The relatively steeply sloping edge of a continental block that lies between the continental shelf and a continental rise or the deep-ocean floor.

convergence   A line or zone where waters of different origins come together; often accompanied by sinking of surface waters.

coral reef   A wave-resistant ridge or mass formed by coral algae and shell fragments of other marine organisms cemented together.

core (sediment)   A vertical, cylindrical sediment sample; may be used to study the age and composition of sediment deposits.

Coriolis effect   The apparent force that deflects a moving object (to the right in the Northern Hemisphere, to the left in the Southern Hemisphere); caused by the earth's rotation.

crest (of a wave)   The highest point of a wave.

crustal plate   A section of the earth's crust that moves independently with respect to other sections in response to movements in the earth's deeper layers.

deep zone   The region below about 2,000 meters (1.25 miles) depth in the ocean; contains approximately 80 percent of ocean waters.

delta   A deposit of riverborne sediment, as at the mouth of an estuary or tidal inlet.

density   The mass per unit volume of a substance, often expressed in grams per cubic centimeter.

diurnal tide   A tide exhibiting one high and one low water per tidal day of 24 hours 25 minutes.

dune   An accumulation of wind-blown sand behind a beach; deposited above the highest point reached by storm waves.

eastern-boundary current   The weak, shallow currents that return waters toward the equator in an open-ocean current gyre.

ebb current   The movement of tidal current away from shore or down a tidal stream.

eddy   A small, nearly closed current system that may move with surrounding waters.

Ekman spiral   Theoretical representation of currents resulting from a steady wind; surface waters move 45 degrees to the right of the wind in the Northern Hemisphere, and water at successive depths moves more slowly and still more to the right, resulting in a water transport that is 90 degrees to the right of the wind.

Ekman transport   A net transport of waters in the wind-driven surface layer that is 90 degrees to the right of the wind in the Northern Hemisphere, 90 degrees to the left of the wind in the Southern Hemisphere.

estuarine circulation   A circulation pattern in which low-salinity surface waters from the river move seaward, entraining the higher-salinity waters in deeper layers so that additional seawater flows into the estuary along the bottom.

estuary   A semienclosed body of water with free connection to the sea; commonly the lower end of a river.

fetch   Area in which waves (seas) are generated by steady winds; also the distance over which the winds act.

flood current   Current associated with a rising tide; directed toward the shore.

forced wave   A wave generated and maintained by a continuous force, in contrast to a free wave, which continues after the generating force ceases.

fringing reef   A reef attached to the shore.

front   A surface separating two dissimilar water masses.

geostrophic current   A current resulting from the balance between gravitational forces and the Coriolis effect.

groin   A low wall of durable material built near the coast and intended to modify sand or water movements.

gulf   A portion of an ocean or sea partly enclosed by land.

gyre   A circular, semiclosed current system.

halocline   A water layer exhibiting marked vertical change in salinity.

heat capacity   The amount of heat required to raise the temperature of a substance by a given amount.

inlet   A narrow passage between islands or a long, narrow indentation of a shoreline.

island arc   A group of volcanic islands usually arc shaped and generally convex toward the ocean where it is associated with a deep trench or trough, and having a deep basin on the concave side.

jetty   A structure built to modify tidal currents or prevent sedimentation of channels, sometimes also to protect the entrance to a harbor or river.

lagoon   A shallow water body separated from the open ocean by a reef or barrier island.

latent heat of evaporation   Heat absorbed per unit mass by evaporation of a liquid.

littoral drift   Sand moved parallel to the shore by waves and currents.

manganese nodules   Lumps of iron and manganese; widely distributed over the ocean floor.

mantle   The relatively plastic zone between the earth's crust and core.

meander   A part of a current that is temporarily diverted to follow its own winding path before merging again with the main current.

mid-oceanic ridge (rise)   A large ridge or ocean-bottom swell extending through an ocean basin, often roughly parallel to continental margins.

mixed tide   A tide exhibiting two unequal high and low waters per tidal day.

neap tide   The tide exhibiting the least range; occurs near the first and third quarters of the moon, when the attraction of the sun and moon do not coincide.

node   The part of a standing wave where vertical motion is least and horizontal water movements are greatest.

North Atlantic Deep Water   A water mass formed in the arctic region that sinks to the bottom of ocean basins at high latitudes in the Northern Hemisphere.

nutrient minerals   Substances required for growth by marine plants; for example, nitrates and phosphates.

ocean basin   Part of the ocean floor that lies below 2,000 meters (1.25 miles).

oceanic crust   Basaltic material, typically 5 kilometers (3 miles) thick, that underlies the oceans.

plume   A distinctive water mass that is identifiable for some distance from its source, for instance a river mouth.

progressive wave   A wave in which the wave form always moves in the same direction.

pycnocline zone An oceanic water layer characterized by sharp vertical density changes; associated with salinity and/or temperature changes with depth.

reef (coral) A wave-resistant structure, often a hazard to navigation, occurring within 20 meters (60 feet) or less of the water surface.

residence time Defined for an element in seawater by the rate of its introduction or removal divided into the total amount of the element present in the water.

reversing tidal current A tidal current that flows alternately in approximately opposite directions, with a period of slack water at each reversal.

rift valley A long, narrow depression bounded by faults, for instance at the summit of and running parallel to the crest of a mountain chain, such as a mid-oceanic ridge.

rotary tidal current A tidal current that flows continuously through all points of the compass with the direction of flow, as in the open ocean where no barriers restrict current flow.

salinity A measure of the amount of salts dissolved in seawater; usually expressed as parts per thousand by weight.

salt-wedge estuary An estuary exhibiting the characteristic estuarine circulation, where a wedge-shaped layer of seawater intrudes into the estuary at the bottom, its salinity decreasing with proximity to the overlying layer of seaward-flowing river water.

sea Irregular, choppy waves in the region where they are generated.

sea-floor spreading The process by which oceanic crust is generated by volcanic action at mid-oceanic ridges and is drawn toward trenches near ocean-basin margins, where it reenters the earth's plastic mantle.

sea-surface topography The low relief in sea level caused by piling up of low-salinity waters, for instance at the centers of current gyres and near coastlines where large amounts of fresh water enter the ocean.

semidiurnal tide A type of tide characterized by two equal high and low waters per tidal day of 24 hours 25 minutes.

shallow-water wave A wave having a wave length more than twice the water depth.

shoreline The boundary between the land and a body of water.

slack water The time when a tidal current has near-zero velocity; usually at the time of a direction reversal.

spring tide A tide of near-maximum tidal range, when the attractions of the sun and moon for ocean water coincide, as at new and full moons.

standing wave A type of wave motion in which the water surface oscillates vertically but does not move horizontally; usually occurs in an enclosed basin, whose period controls the wave period.

storm surge A high water level against the shore; associated with strong winds and sometimes with unusually high tides.

surface zone The uppermost layer of ocean water, usually 200 meters (600 feet) or less deep, where seasonal temperature and salinity changes govern water properties and wind mixing homogenizes the water layer vertically.

surf zone The shallow-water zone where waves break in a coastal area; between the outermost breaker and the limit of wave uprush on the beach.

swell Smooth, regular waves that have traveled out of the area where they were generated.

temperature of initial freezing   The temperature at which seawater begins to freeze; decreases with increased salinity.

thermocline   A zone where water temperatures sharply decrease with depth, as in an ocean-water layer separating a vertically homogeneous surface zone from the deep, constant-temperature zone below.

tidal bulge (tidal crest)   A long-period wave associated with the tide-producing forces of the sun and moon.

tidal current   A horizontal water movement associated with the rise and fall of the tide in response to tide-producing forces.

tidal marsh   A swampy area which is covered by water during high tides but is exposed at low tide.

tidal range   The difference in height between high and low waters.

tide rip   An agitation of water caused by the meeting of opposing currents.

transform fault   A fracture in the rocks between discontinuous sections of a crustal plate.

trench   A long, deep, narrow depression in the ocean floor; especially an area where oceanic crust is drawn into the earth's mantle.

trough (of a wave)   The lowest part of the wave form.

upwelling   The process by which deep waters rise to the surface; often prominent where a wind-driven current carries coastal water seaward.

water mass   A body of water identifiable within a larger body by its distinctive chemical and/or physical properties.

wave height   The distance between the trough and crest of a wave.

wave length   The distance between successive wave crests or troughs.

wave period   The time between the passage of successive wave crests or troughs at a given location.

western-boundary current   Typically, a fast, deep current that carries water from an equatorial current toward higher latitudes in an open-ocean current gyre.

general bibliography

Dietrich, Gunter. *General Oceanography: An Introduction.* Translated by Feodor Ostapoff. New York: Interscience, 1964. *General reference. Good technical bibliography.*

Gross, M. Grant. *Oceanography.* Englewood Cliffs, New Jersey: Prentice-Hall, 1972. *College-level introductory text.*

Schlee, Susan. *The Edge of an Unfamiliar World: A History of Oceanography.* New York: E.P. Dutton & Co., 1973. *Covers oceanography in the nineteenth and twentieth centuries.*

Sverdrup, H.U., Johnson, M.W., and Fleming, R.H. *The Oceans: Their Physics, Chemistry, and General Biology.* New York: Prentice-Hall, 1942, 1970. *A classic in the field.*

Thurman, Harold V. *Introductory Oceanography.* Columbus, Ohio: Charles E. Merrill Publishing Company, 1975. *Comprehensive, general text.*

von Arx, W.S. *An Introduction to Physical Oceanography.* Reading, Massachusetts: Addison-Wesley, 1962. *Intermediate difficulty.*

index

References to book pages are given in lightface type, followed by references to MEDIAPAK frame numbers, indicated in boldface.